WITHDRAWN

◆ *The Emergent Self*

BOOKS BY WILLIAM HASKER

The Emergent Self

God, Time, and Knowledge

Metaphysics: Constructing a Worldview

The Openness of God: A Biblical Challenge to the Traditional Understanding of God, with Clark Pinnock, Richard Rice, John Sanders, and David Basinger

Philosophy of Religion: Selected Readings, edited with Michael Peterson, David Basinger, and Bruce Reichenbach

Reason and Religious Belief: An Introduction to the Philosophy of Religion, with Michael Peterson, David Basinger, and Bruce Reichenbach

The Emergent Self

✦ WILLIAM HASKER

Cornell University Press ITHACA AND LONDON

First published 1999 by Cornell University Press

Printed in the United States of America

LIBRARY OF CONGRESS CATALOGING-IN-PUBLICATION DATA

Hasker, William, 1935–
    The emergent self / William Hasker.
        p.   cm.
    Includes index.
    ISBN 0-8014-3652-4 (alk. paper)
      1. Philosophy of mind.   2. Mind and body.   I. Title.
    BD418.3.H375   1999   99-26814
    128'.2—dc21

Cornell University Press strives to use environmentally responsible suppliers and
materials to the fullest extent possible in the publishing of its books. Such materials
include vegetable-based, low-VOC inks, and acid-free papers that are recycled,
totally chlorine-free, or partly composed of nonwood fibers. Books that bear the logo of
the FSC (Forest Stewardship Council) use paper taken from forests that have been
inspected and certified as meeting the highest standards for environmental and social
responsibility. For further information, visit our website at www.cornellpress.cornell.edu.

Cloth printing   10   9   8   7   6   5   4   3   2   1

FSC   FSC Trademark © 1996 Forest Stewardship Council A.C.
        SW-COC-098

FOR BENJAMIN
an emerging self

You do not know how the spirit comes to the bones in the womb of a woman with child.

—Ecclesiastes 11:5

Contents

Preface

This book sets out to carve a path through the jungle that is contemporary philosophy of mind. To refer to the philosophy of mind as a jungle implies no disparagement. The field is marked by the exuberant growth and rapid evolution of an astonishing diversity of theories and perspectives, and is so fiercely competitive that even the most vigorous specimens must struggle in order to survive. The richness and complexity of this branch of philosophy make it difficult to get a clear overview of the whole, let alone master it. Broad surveys of the territory are like satellite images: useful in their own way, but not all that helpful to the wayfarer slogging along on the ground. But it could hardly be otherwise, given the broad range of topics and problems comprised in the field: one is reminded of Wilfrid Sellars's remark that the mind-body problem "soon turns out, as one picks at it, to be nothing more nor less than the philosophical enterprise as a whole." That may be hyperbolic, but the point Sellars was making can't be denied. No attempt will be made here to take on the whole enterprise. Important topics are given little or no attention (for example, the problem of mental content), and the material covered in any one of the chapters could be expanded to fill an entire volume. The overall aim is to present and defend a particular stance on the mind-body problem, a stance here termed "emergent dualism." This position is not without precedent, but neither is it at all common; if readers come to perceive it as a viable option along with the more familiar candidates, my efforts will be amply rewarded.

This particular path through the jungle is charted to pass midway between the warring camps of materialists and dualists. My strategy has been compared by a colleague to that of a Civil War rifleman who went into battle wearing a blue shirt and gray pants. If he expected to avoid hostile fire, presumably he was disappointed. I have no such expectation. I do hope, however, that at least a few of the combatants may come to see in my proposals the basis for a negotiated settlement. But there is one kind of approach to these issues that is unlikely to be affected by the views and arguments contained in this book. As an example of this approach (though by no means the only one) we may take Daniel Dennett, as he presents himself in his essay in *A Companion to the Philosophy of Mind* (Blackwell, 1995). He tells us that, having come to distrust the methods employed by other philosophers, he decided that "before I could trust any of my intuitions about the mind, I had to figure out how the brain could possibly accomplish the mind's work." This means accepting, right from the outset, that the brain is a "syntactic engine" that mimics the competence of "semantic engines." (How we mere syntactic engines could ever know what a semantic engine might be is not addressed.) All this is dictated by an "initial allegiance . . . to the physical sciences and the third-person point of view," an allegiance which in turn is justified by appeal to an evolutionary perspective. The foundational commitment to a mechanistic materialism is unmistakable. This commitment is subsequently refined and elaborated, but it is never subjected to a fundamental reevaluation; rather, data that conflict with it are dismissed as illusory. ("This conviction that *I*, on the inside, deal directly with meanings turns out to be something rather like a benign 'user illusion.'") In view of this, it seems appropriate to characterize Dennett's physicalism as a *dogmatic presupposition*—and such dogmatism is hardly rendered benign by the fact that it is fairly widespread in the philosophy-of-mind community. Philosophers determined to hold on to a dogmatic materialism at all costs will hardly be influenced by anything said here to the contrary. It is my hope, however, that a good many readers, including both materialists and nonmaterialists, will find the arguments and analyses in these pages helpful as they subject *both* traditional dualisms and the multifarious modern materialisms to severe scrutiny.

This book could not have been written without a lot of help from my friends. After some thought I have concluded that too many people have made significant contributions for all of them to be mentioned in this preface. Three men whose help has been especially important at various stages are Timothy O'Connor, Victor Reppert, and Dean Zimmerman. To the many other friends and colleagues whose contributions have not been individually acknowledged here, I offer this consolation: at least you have escaped being associated with my mistakes! None of the mistakes are due to Huntington College, which granted me a sabbatical, nor to the Pew Evangelical Scholars Program, whose provision of a research fellowship enabled me to spend a year in intensive study and writing. I must acknowledge, also, the helpfulness of John Ackerman and the editors at Cornell University Press. And finally, my thanks to Charles Taliaferro, who read the manuscript for the Press and made a number of helpful comments; any stylistic infelicities that remain are almost certainly due to my failure to take his suggestions.

✦ *The Emergent Self*

CHAPTER ONE

What Can't Be Eliminated

Let us begin with a modest proposal: there are intentional conscious experiences. There are, that is to say, such episodes as a person wondering whether it is going to rain, or believing that this has been an unusually cold winter, or deciding to let the credit card balance ride for another month. In typical cases such as these the *intentional content* of the experience, what the experience is about, is something distinct from the experience itself, something that could exist or obtain (or fail to exist or obtain) regardless of whether or not the experience occurred. These episodes are *consciously experienced*; when we have them we are aware that we are having them, and there is "something it is like" to be having them.

The claims in the preceding paragraph are extremely modest compared with other, more expansive claims that have been made about these topics. I have not asserted, with Brentano, that "intentionality is the hallmark of the mental"; it is possible that there may in fact be conscious, mental experiences which are not intentional. Devotees of Transcendental Meditation claim to reach a state of "pure awareness" which lacks all content; more prosaically, there are mood-states which should be classified as mental but seem to lack any discernible intentional content. Claims such as these, purporting to identify mental states which are not intentional, possess some initial plausibility, but there is no need to decide at this point whether to accept them or not.

It has not been asserted that all intentional states are conscious, and this would seem not to be true: believings, hopings, intendings, and the like, are in many instances *dispositional* states that can endure throughout lengthy periods during which they are not consciously experienced or attended to. Nor have we assented to John Searle's more plausible claim that all intentional states are at least potentially and in principle accessible to consciousness.[1] There is much to be said for this, but at this point we need not commit ourselves. And finally, we have not endorsed Sartre's assertion that "consciousness is consciousness of itself."[2] Sartre's claim threatens to generate infinite complexity within each and every conscious state: if being aware of X includes being aware of being aware of X, does it also include being aware of being aware of being aware . . . ? But once again, there is no need to decide for or against Sartre at this juncture.

So in asserting that there are intentional conscious states we have avoided the controversial "add-ons" that might seem to make this a contentious, and therefore interesting, philosophical assertion. Why, then, bother to make a point of something so trivial and self-evident? But these are not ordinary times in philosophy, and there are plenty of authorities who make it plain to us that even such a modest assertion as this one will not pass without a fight. So we must give some attention to eliminative materialism.

ELIMINATIVE MATERIALISM

As the starting point for our discussion of eliminative materialism, we take Paul Churchland's now-classic 1981 paper, "Eliminative Materialism and the Propositional Attitudes."[3] Here he states, "Eliminative materialism is the thesis that our commonsense conception of psychological phenomena constitutes a radically false theory, a theory so fundamentally

1. See John Searle, *The Rediscovery of the Mind* (Cambridge: MIT Press, 1992), chap. 7.

2. Jean-Paul Sartre, *The Transcendence of the Ego: An Existentialist Theory of Consciousness*, trans. Forrest Williams and Robert Kirkpatrick (New York: Noonday Press, 1957), p. 40.

3. Churchland, "Eliminative Materialism and the Propositional Attitudes," *Journal of Philosophy* 78, no. 2 (1981); reprinted in Paul Churchland, *A Neurocomputational Perspective: The Nature of Mind and the Structure of Science* (Cambridge: MIT Press, 1989), pp. 1–22. Page references to Churchland, unless otherwise noted, are to this volume.

defective that both the principles and the ontology of that theory will eventually be displaced, rather than smoothly reduced, by completed neuroscience" (p. 1). In support of this Churchland argues, first of all, that our commonsense conception of the mental, dubbed "Folk Psychology," constitutes a *theory* for the explanation and prediction of behavior, a theory which, like other theories, is in principle open to replacement in view of the advance of science. The meanings of the terms in this theory—in this case, especially the terms referring to such propositional attitudes as belief, desire, and the like—are "fixed or constituted by the network of laws" (p. 3) in which they occur—and since the laws are "radically false," there is no escaping the conclusion that the entities ostensibly referred to by the terms in question *do not exist*; they are in the same category with ether, phlogiston, caloric fluid, and vital spirits (all Churchland's own examples).

But, granted the theoretical status of folk psychology, why should we suppose that its being false is really a live option, as eliminative materialism claims? Should we not, on the contrary, take the view that its integrity is guaranteed by the "substantial amount of explanatory and predictive success" the theory admittedly enjoys? Churchland thinks not. He points to the numerous explanatory and predictive *failures* of folk psychology, as seen (for example) in "the nature and dynamics of mental illness, the faculty of creative imagination, or the ground of intelligence differences between individuals." Furthermore, the history of folk psychology reveals a story of "retreat, infertility, and decadence"; it constitutes, in Lakatosian terms, a degenerating research program. Finally (and perhaps most important) folk psychology contrasts most unfavorably with the "coherent story of the species' constitution, development, and behavioral capacities that encompasses particle physics, atomic and molecular theory, organic chemistry, evolutionary theory, biology, physiology, and materialistic neuroscience"—a growing synthesis in which folk psychology plays no part, and into which it apparently cannot be integrated (pp. 6–9).

In his slightly later *Matter and Consciousness*,[4] Churchland makes clear that we are not necessarily limited to the alternatives of keeping folk psychology as it stands or replacing it with something radically different. The main point at issue is how the categories of our commonsense framework will match up with "the correct neuroscientific account of human capaci-

4. Paul Churchland, *Matter and Consciousness: A Contemporary Introduction to the Philosophy of Mind* (Cambridge: MIT Press, 1985).

ties." But this matching is in principle a matter of degree; pure reduction and pure elimination "are the end points of a smooth spectrum of possible outcomes, between which there are mixed cases of partial elimination and partial reduction."[5] Nevertheless, eliminative materialism as such must be taken to be a view which, at the least, "lies substantially toward the revolutionary end of the spectrum."

The case sketched out for eliminative materialism has been subjected to attack at many different points. The claim that our commonsense conception of the mental constitutes a "theory" that is in competition with, and replaceable by, scientific theories has been criticized extensively (and, I think, effectively) by Lynne Rudder Baker.[6] Terence Horgan and James Woodward, among others, have argued that the practical success of folk psychology in everyday life, and its evolutionary success in the history of our species, are such that it would be astonishing if it turned out to be fundamentally false.[7] John Dupre charges the eliminativists with "scientistic arrogance," revealed in their assumption that folk psychology can be replaced by neurobiology.[8] And most users of the commonsense conception would be bewildered to learn that they were engaged in a "research program," to be evaluated by its success in generating scientific progress. On the other hand, if folk psychology is to be viewed as a research program, the verdict that it is "non-progressive" requires one to subscribe to the rather remarkable judgment that *none of the existing social sciences* (all of which operate within this framework)[9] *have made significant progress.*

5. Ibid., p. 49.

6. See Baker, "The Myth of Folk Psychology," in *Explaining Attitudes: A Practical Approach to the Mind* (Cambridge: Cambridge University Press, 1995).

7. Terence Horgan and James Woodward, "Folk Psychology Is Here to Stay," *Philosophical Review* 94 (1985): 197–220.

8. "Folk psychology provides us with the entire resources of language for making subtle distinctions between an indefinite array of possible mental states. This is one of the main things that language is for, and one of the things that language users, especially the most talented users, are most skilled at applying it to. One should be astounded and even appalled at the scientistic arrogance that supposes that the dry and highly specialized technical terminology of an esoteric subfield of science should supplant the instrument developed over centuries by the efforts of Shakespeare, Dante, and Dostoevsky—to speak only of the West—and many millions of others to describe the subtleties of the human mind" (Dupre, *The Disorder of Things: Metaphysical Foundations of the Disunity of Science* [Cambridge: Harvard University Press, 1993], p. 158).

9. Behaviorism in psychology may be an exception, though it has been argued that the descriptions given of behavior are in fact parasitical on folk-psychological no-

Yet another line of attack is found in the suggestion of Owen Flanagan that, even if folk psychology is inadequate as an *explanatory theory*, the conscious experiences noted by folk psychology remain as *data* that any more adequate theory will need to take account of.[10] The eliminativist's response to this, of course, is that the distinction between theoretical and observational entities is itself a piece of bad theory. In the end, there is no such distinction to be made, and if the theory fails its ontology dies along with it. But whatever the ultimate status of the distinction may be, there certainly *seems* to be a relevant difference between turnips and neural activation vectors—and beliefs and desires, as we experience them in everyday life, seem much more like the former than the latter.

All these challenges to eliminative materialism are substantive and deserve further elaboration. But the most intriguing challenge, and also the most controversial, is found in the claim that eliminative materialism is "self-defeating." To this we now turn.

THE SELF-REFUTATION ARGUMENT

The idea that eliminative materialism is somehow self-defeating or self-referentially incoherent has probably occurred independently to a number of people; Churchland notes that it surfaced in a question from the audience at the very first public presentation of his 1981 paper.[11] Opponents of eliminativism often see this as the final, devastating refutation of this theory, whereas defenders invariably charge that the objection begs the question against eliminativism by assuming the very folk psychology whose integrity is at issue. I shall argue that while some versions of the ob-

---

tions. But for the social sciences generally, consider the following from Stephen Stich: "Economics, political science, sociology, and anthropology are up to their ears in the intentional idiom that is the hallmark of folk psychology. If all talk of beliefs, desires, expectations, preferences, fears, suspicions, plans, and the like were banished from the social sciences, those disciplines as we know them today would disappear" (*From Folk Psychology to Cognitive Science: The Case against Belief* [Cambridge: MIT Press, 1983], p. 213).

10. Owen Flanagan, *Consciousness Reconsidered* (Cambridge: MIT Press, 1992), pp. 26–27.

11. Churchland, "Postscript: Evaluating Our Self Conception," to "Eliminative Materialism and the Propositional Attitudes," in Paul K. Moser and J. D. Trout, eds., *Contemporary Materialism: A Reader* (London: Routledge, 1995), pp. 168–79; the remark cited is on p. 170.

jection are indeed question-begging, it nevertheless holds the potential to do more damage to eliminativism than its defenders have been willing to admit.

The most extensive development of this objection is by Lynne Rudder Baker, who argues that eliminative materialism faces "the threat of cognitive suicide."[12] The reason for this is that "if the thesis denying the common-sense conception of the mental is true, then the concepts of rational acceptability, of assertion, of cognitive error, even of truth and falsity are called into question" (p. 134). To support her claim she examines in some detail the notions of rational acceptability, assertion, and truth, arguing in each case that our existing concepts presuppose the notions of propositional attitudes and of mental content expressible in "that"-clauses. (She does acknowledge [p. 143] that if one endorses a redundancy theory of truth, the problems about truth would reduce to those about assertibility.) In view of this, she concludes, "it seems that we can neither rationally accept nor assert nor even formulate the thesis denying the common-sense conception of the mental. Indeed, if the thesis is true, it is at least problematic whether we can rationally accept or assert or even formulate any thesis at all. This seems ample reason to deny the conclusions of the argument from physicalism" (p. 147).

Baker's detailed argument is too long to reproduce here, but as a representative sample, consider some of her remarks about assertion. Her argument, she says, "challenges the eliminative materialist to show how there can be assertion without belief or other states with content" (p. 139). The argument does not presuppose any elaborated theory of meaning; it only "makes the minimal assumption that language can be meaningful only if it is possible that someone mean something" (p. 140). She then goes on to specify three requirements that would have to be met by an account of assertion that might be proffered by the eliminativist:

(i) Without appeal to the content of mental states, the alternative account of assertion must distinguish assertion from other audible emission.

She notes that such an account might "distinguish between kinds of causal history. But it is difficult to guess how to specify the right causal

12. This is the title of chapter 6 of Baker's *Saving Belief: A Critique of Physicalism* (Princeton: Princeton University Press, 1987). Page references to Baker, unless otherwise noted, are to this work.

history without attributing to the speaker some state with the content of what is asserted." Continuing,

> (ii) The alternative account of assertion, again without appeal to the content of mental states, must distinguish sounds that count as an assertion that *p* rather than as an assertion that *q*.
>
> (iii) The alternative account of assertion must at least have conceptual room for a distinction between sincere assertion and lying. (p. 141)

Since it seems dubious that any account proffered by the eliminativist can meet these requirements, she concludes, "I think we have substantial reason to doubt that any alternative account of assertion that is free of appeal to contentful mental states will be forthcoming" (pp. 141–42).

How might an eliminativist respond to this argument? One possibility would be to object that the argument is question-begging merely in virtue of its being stated in terms of the folk psychology the eliminativist rejects. This, however, is a very costly response for the eliminativist to make. If he is unable or unwilling to respond to philosophical comments that employ our commonsense conception of the mental, then (in the lack of any serviceable alternative vocabulary) he is condemned to virtual silence in philosophical discussion. For a philosopher, at any rate, this does indeed seem a short route to "cognitive suicide."

Another response would be to deny, in spite of Baker's arguments, that our ordinary concept of assertion presupposes propositional-attitude psychology. This denial, however, is greatly lacking in plausibility. It seems to virtually all of us that, for example, the difference between genuine assertion and playacting is that in one case, but not in the other, one intends to convey to the hearer that one is uttering sentences one believes to express truths. To deny this is to say in effect that while we do have the concept of assertion, we are completely mistaken about what the implications of the concept are. In the absence of any plausible alternative analysis it is difficult to take this seriously.[13]

A third, and somewhat more promising, response, is to say that whereas our ordinary concept of assertion presupposes propositional-atti-

---

13. Here we should note Stephen Stich's observation that it is difficult to see how the notion of a "sincere assertion of p" could be "unpacked without invoking the idea of an utterance *caused by the belief that p*" (*From Folk Psychology to Cognitive Science*, p. 79; emphasis in original).

tude psychology, an adequate account of the actual phenomena we now designate as "asserting" will employ a "successor concept" which does not have this objectionable presupposition. This successor concept, it will be admitted, has yet to be developed, but when it makes its appearance the illusion that belief, desire, and the rest are real, existing mental states will have been dispelled.

This response still faces formidable difficulties. The three requirements Baker puts forward on an alternative account of assertion might plausibly be taken also as requirements on any successor concept that will do (approximately) the same work now done by the concept of assertion—and if so, the prospects still seem bleak. On the other hand, the requirements for a "successor concept" are presumably somewhat looser than those for an analysis of a concept already in common use, so there is at least a bit more room for maneuver than in the previous case. On the whole, it would seem that this third response is the most promising one for the eliminativist to adopt. And if some obstacles remain, what of it? Major scientific breakthroughs frequently undermine assumptions that, prior to the breakthrough, seemed beyond challenge, so why should this case be different?

DOES THE SELF-REFUTATION ARGUMENT BEG THE QUESTION?

Is the self-refutation argument against eliminativism question-begging? In order to answer this, it will be helpful to begin with some arguments that are simpler, and more compactly stated, than Baker's. Consider then the following "naive transcendental argument" devised by Michael Devitt:

1. The eliminativist sincerely utters, "There are no beliefs."
2. So, the eliminativist believes that there are no beliefs.
3. So, eliminativism about beliefs involves realism about beliefs.
4. So, eliminativism is incoherent.

Clearly this *is* question-begging; as Devitt observes, the argument "*starts by ignoring what the eliminativist actually says.* Since she *is* an eliminativist, she rejects the established intentional way of talking. So she will not describe any mental state, *including her own in stating eliminativism,* as a belief. So step 2, which saddles her with precisely what she is denying, is blatantly question-begging."[14]

But not all self-refutation arguments are as naive as this one. Consider next an argument adapted from one constructed by Victor Reppert:

1. Either eliminative materialism has been meaningfully asserted, or it has not been meaningfully asserted.
2. If eliminative materialism has been meaningfully asserted, then the assertion was produced by someone who has the belief that eliminative materialism is true.
3. If the assertion of eliminative materialism was produced by someone who has the belief that eliminative materialism is true, then there are beliefs and eliminative materialism is false.
4. If eliminative materialism has not been meaningfully asserted, then eliminative materialism has not been made publicly intelligible.
5. Therefore, either eliminative materialism is false, or eliminative materialism has not been made publicly intelligible.[15]

Is this argument question-begging? In answering this, Reppert does something omitted by many of those who discuss the topic: he investigates with some care exactly what constitutes a question-begging argument. In the simplest case, an argument begs the question if its conclusion appears among its premises. This presumably happens rather seldom, and then only by inadvertence. But an argument may also be question-begging if it *employs premises that rely on the conclusion for support.* This, however, may not always be obvious, since what support someone has for the premises of an argument she endorses is often not apparent. To get around this, Reppert advocates the "principle of charity": an argument should be adjudged question-begging only if "*no* reasonably well-informed person would accept the premise who does not already accept the conclusion."[16] As an example of an argument which is clearly question-begging by this

14. Michael Devitt, "Transcendentalism about Content," *Pacific Philosophical Quarterly* 71 (1990): 247–63; the quotation is from p. 248.

15. See Victor Reppert, "Eliminative Materialism, Cognitive Suicide, and Begging the Question," *Metaphilosophy* 23 (1992): 378–92; for the argument (here slightly abbreviated), see p. 380. It should be noted that this argument is put forward by Reppert merely as an example for his discussion of begging the question.

16. Ibid., p. 389.

standard, Reppert cites the case of someone who would prove that God exists because the Bible, which contains no error, says that God exists. All parties to the discussion, he thinks, would recognize that the only reasonable ground for claiming the Bible to be inerrant is that it was inspired by God, so the argument assumes what it set out to prove. If on the other hand the premises of an argument are such that a reasonable, well-informed person might have support for them that is not based on the acceptance of the conclusion, then the argument is not question-begging, whatever other faults it may suffer from.

Now, how does this analysis apply to the previously stated argument? The premises of the argument most likely to lay it open to the charge of begging the question are

2. If eliminative materialism has been meaningfully asserted, then the assertion was produced by someone who has the belief that eliminative materialism is true,

and perhaps also

4. If eliminative materialism has not been meaningfully asserted, then eliminative materialism has not been made publicly intelligible.

But is it really the case that (2) and (4) could not be accepted by any reasonable, well-informed person who had not already accepted the conclusion of the argument? Premise (2) is presumably an inference from

2'. If *anything* has been meaningfully asserted, then the assertion was produced by someone who has the belief that what was asserted is true.

But of course, all manner of persons have accepted premise 2' long before they had ever heard of eliminative materialism, so it can hardly depend for its support on some conclusion about the latter. Rather, it may be thought to be supported by the social entrenchment and the very great practical usefulness of our ordinary conception of the mental, embedded as it is in a vast array of our everyday cognitive practices.[17] And even after she has learned about eliminativism, someone might well conclude that

17. My line of argument here differs somewhat from Reppert's, but the overall effect is the same.

the arguments in its favor, which are acknowledged even by its supporters to be less than conclusive, cannot outweigh the massive support which 2' enjoys for her. Similar remarks, mutatis mutandis, could be made about the support for premise (4). So, Reppert concludes, the argument he has given does *not* beg the question.

Reppert's analysis of begging the question is extremely helpful, but I believe it requires supplementation to meet the special case of a *self-refutation* argument. As he rightly says, "An argument might be regarded as an invitation to make an inference."[18] And this naturally gives rise to the question, *Who* is being invited to make an inference? On Reppert's account of begging the question, the answer is that this doesn't matter. But for an argument to show that a given position is *self*-refuting, one must "invite" (or, if possible, compel!) *the proponent of that position* to make an inference that ends up undermining the position. And in order to do this, one must appeal to premises that *are or should be accepted by the proponent of the position in question.* Now in order to apply this to Reppert's argument, we need to determine more accurately the position of the eliminativist who is its target. If she is what might be termed a "naive" eliminativist, one who has assumed without much reflection that she could abandon belief but continue to employ the concept of assertion, then the argument may well succeed in showing that her position is self-refuting. For by calling attention to the logical connection between assertion and belief (for present purposes, I assume that this connection does indeed obtain), it forces her to accept the unwelcome conclusion that her eliminativism either is false or has not been made publicly intelligible.

But what about a more sophisticated eliminativist, of the kind discussed in the preceding section? This sophisticated eliminativist recognizes the link between assertion and belief, so she will readily grant premise (2) of Reppert's argument. She will not, then, consider that she has *asserted* eliminative materialism, but she clearly thinks she has somehow put that theory forward for public consideration *without* having asserted it. (Presumably, what she has done would be accurately described by the "successor concept" to assertion in the developed eliminativist framework—but this successor concept is not, as yet, available for analysis.) So the denial of (4) is an integral part of the position of the sophisticated eliminativist. And in inviting her to accept (4) as a premise, Reppert's argument in effect invites her to assume that her position is false, in

18. Reppert, "Eliminative Materialism," p. 388.

order to conclude that it is self-refuting. In this context, the argument may well be judged guilty of begging the question. To be sure, the argument might *not* be question-begging, if given by the non-eliminativist *as a reason for his own rejection of eliminativism*. And on the other hand, the argument might, without begging any questions, be successful in persuading someone not to become an eliminativist in the first place. But taken in either of these ways, it would not be a *self*-refutation argument.[19]

And now, finally, what of Baker's argument? It is important to note that she (unlike Reppert and Devitt's "naive transcendentalist") explicitly considers the possibility that the eliminativist may be able to develop alternative concepts which will do the work, in the eliminativist's scheme, now done for us by the concepts of rational acceptance, assertion, and truth. "It remains to be seen," Baker writes, "whether such concepts (or suitable successors) can be constructed" within the constraints imposed by eliminative materialism. As we shall see, this has an important bearing on the issue of begging the question.

Baker does not explicitly set out her argument in a series of numbered steps, and its actual structure is more complex than might appear. I believe its main contentions can be summarized as follows:

A. Our present concepts of rational acceptance, assertion, and truth are incompatible with eliminative materialism.
B. Unless it is possible for eliminative materialism to develop suitable successor concepts to rational acceptance, assertion, and truth, eliminative materialism is self-refuting.
C. Probably, eliminative materialism cannot develop such successor concepts.
D. Probably, eliminative materialism is self-refuting.

Is this argument question-begging? (A) is strongly supported by evidence which is neutral territory among the parties to the dispute; namely by the logical facts about our present concepts, and the restrictions imposed by eliminative materialism. Furthermore, (A) would be accepted by a good many eliminativists. If (A) is accepted, (B) seems undeniable—and once

---

19. In discussion, Reppert has said that he does not think the criteria for begging the question should be different for self-refutation arguments than for others. He does agree, however, that we should not expect the type of argument he presents to be convincing to an eliminativist.

again, most eliminativists would agree. We can expect the eliminativist to object to (C). But here it is important to note that (C) *is not introduced as a premise; rather, it is supported by argument.* The premises of this argument are essentially the same as the evidence for (A): logical facts about our present concepts, and the roles they play in our cognitive economy, and the restrictions on successor concepts imposed by eliminative materialism. And the argument from these premises to (C), which is persuasive though not deductively valid, does not rely on any special assumptions the eliminativist might reject. And finally, (D) follows straightforwardly from (B) and (C). In all of this, there is nothing the eliminativist would reject merely in virtue of being an eliminativist. *The charge of begging the question cannot be sustained against Baker's argument.*[20]

IS ELIMINATIVE MATERIALISM SELF-REFUTING?

So the argument is not question-begging, but is it successful? Here a great deal depends on the possibility of the eliminativist developing successor concepts to the concepts which figure in our commonsense conception of the mental. Baker has argued, quite effectively I think, that this cannot be done. But her arguments do not *conclusively establish* the impossibility, nor does she claim to have done this. Arguments of this sort characteristically observe that "it is difficult to guess" how such-and-such could be done, that "it is difficult to see how anything could count" as having succeeded in the task, and so on. But such remarks always leave open the retort that the fault lies in the defective imagination of the one who can't see how something could be done. And on the other hand, a stronger argument, one that would establish conclusively the impossibil-

20. Rod Bertolet does charge Baker with begging the question, on the ground that whether or not eliminativism is tenable depends on the future development of science, which we are not now in a position to predict. (See his "Saving Eliminativism, " *Philosophical Psychology* 7, no. 1 [1994]: 87–100.) The difference between us hinges largely on the fact that he attributes stronger claims to Baker than I do. He interprets her as claiming to prove conclusively that it is impossible for satisfactory successor concepts to those of current folk psychology to be created. ("She is quite right that it is more than a little difficult to see how the replacement might go. This is not good enough. What Baker must show is that there is no way it can go " [p. 95].) Baker may very well *believe* that such a replacement is impossible, but I do not think she claims to have *proved* that it is.

ity in question, would almost certainly have to employ premises the eliminativist would reject. I conclude, tentatively, that it has not been shown that the construction of appropriate successor concepts is impossible.

One should, however, take with handfuls of salt the scenarios Churchland has proposed in seeking to overcome our resistance to the idea of such a replacement. What he does is construct science-fictional futures in which our present conception of the mental and of cognitive processes has been replaced by something quite different. One example is as follows:

> Research into the neural structures that fund the organization and processing of perceptual information reveals that they are capable of administering a great variety of complex tasks, some of them showing a complexity far in excess of that shown by natural language. . . .
>
> Guided by our new understanding of these internal structures, we manage to construct a new system of verbal communication entirely distinct from natural language, with a new and more powerful combinatorial grammar over novel elements forming novel combinations with exotic properties. The compounded strings of this alternative system—call them "*Übersätze*"—are not evaluated as true or false, nor are the relations between them remotely analogous to the relations of entailment, etc., that hold between sentences. They display a different organization and manifest different virtues.
>
> Once constructed, this "language" proves to be learnable; it has the power projected, and in two generations it has swept the planet. Everyone uses the new system. The syntactic forms and semantic categories of so-called "natural" language disappear entirely. And with them disappear the propositional attitudes of FP, displaced by a more revealing scheme in which (of course) "*übersätzenal* attitudes" play the leading role. FP . . . suffers elimination.[21]

The verve and enthusiasm with which this little story is told should not obscure for us the fact that we have *no understanding whatever* of the concepts which, in the story, replace our present notions of belief, desire, entailment, rationality, and the like. Churchland's tale may pique the imagination, but philosophically it is of little help.

In subsequent writings, to be sure, Churchland has attempted to spell

21. Churchland, "Eliminative Materialism and the Propositional Attitudes," p. 19f.

out in a somewhat more realistic fashion the nature of a possible successor theory to folk psychology, most notably in his "PDP approach" to explanation and theories.[22] But it remains true that, whereas he alludes to the need for successor concepts to our present notions of (for example) rational evaluation, his discussion sheds little or no light on the actual nature of these successor concepts. The only exception to this generalization I am aware of occurs in a discussion with Bas van Fraassen, where Churchland suggests that we should "move away from the more naive formulations of scientific realism . . . in the direction of *pragmatism*."[23] In the event of such a shift, "Truth, as currently conceived, might cease to be an aim of science."[24] The new theory, as Churchland conceives it, would still qualify as a form of scientific realism in that (1) it shuns the distinction between observables and unobservables characteristic of classic instrumentalism (and of van Fraassen's theory); (2) it retains the idea that "there is a world, independent of our cognition, with which we interact"; and (3) "our best and most penetrating grasp of the real is still held to reside in the representations provided by our best theories."[25]

Whatever its attractions on other grounds, pragmatism holds little promise for eliminative materialism. Eliminative materialism owes much of whatever attraction it possesses to the "naive scientific realism" Churchland proposes to abandon—that is, to the view that science aspires to show us the *real structure of the objective world*, and our best present-day science is at least roughly successful in doing this. To the extent that this is compromised, the pressure to explain everything within the framework of the physical sciences is lessened—but without this, why be materialists at all, let alone eliminativists?

In fact, on any reasonable pragmatic approach the overwhelming practical usefulness of commonsense psychology—whatever its real or imagined limitations—argues strongly against any thought of its elimination. Absent a strong ("naive") scientific realism, the merits of handling all phenomena within a single explanatory framework are merely aesthetic, and not at all compelling in the face of the great pragmatic efficacy of folk psychology. Reppert puts the case well: "The dethronement of truth opens

22. See Churchland, *Neurocomputational Perspective*, chaps. 9 and 10. ("PDP" refers, of course to "parallel distributed processing.")

23. Churchland, "The Ontological Status of Observables," in *Neurocomputational Perspective*, pp. 139–51; the quotation is from p. 149f.

24. Ibid., p. 150.

25. Ibid., p. 151.

up the possibility of a much looser form of pragmatism: a non-propositionalist cognitive science may be the best way to go, propositional attitude attributions are perfectly justified in other contexts, and there just isn't any question of limning the true and ultimate structure of reality. This may not be acceptable to eliminativists like Churchland, but one would like to know why not."[26]

I conclude, then, that no promising successor concepts to the abandoned concepts of folk psychology are visible on the horizon. On the other hand, it has not been proved conclusively that the construction of such successor concepts is impossible, so the eliminativist is seemingly within his epistemic rights to hold on to that possibility and to counsel patience to the rest of us. But does this really leave the eliminativist in a tenable position? To investigate this, I suggest we take a closer look at the actual discourse by which eliminativists commend their theory. Consider then, the following locutions, gleaned from the first two pages of a 1988 article by Churchland:[27]

"our overall epistemic adventure contains both greater peril, and greater promise, than we might have thought"

"The motivation for such a view is not purely philosophical"

". . . experiments designed to illustrate both the inevitable ambiguity of perceptual situations and the cunning resolution of those ambiguities"

"the counterclaim is that . . ."

"My principal aim in this chapter is to show that Fodor's specific claims . . . are mostly irrelevant"

"There are three principal ways in which any perceptual belief may fail of theoretical neutrality"

Each and every one of these locutions either affirms outright or clearly entails the existence of the propositional attitudes eliminativism denies. Each of them, therefore, is in contradiction with the eliminativist's assertion that folk psychology is a "radically false theory." And since it is the latter assertion that must be taken as privileged in this context, Church-

26. Victor Reppert, "Ramsey on Eliminativism and Self-refutation," *Inquiry* 34 (1991): 499–508; the quotation is from p. 507.

27. Churchland, "Perceptual Plasticity and Theoretical Neutrality: A Reply to Jerry Fodor," *Philosophy of Science* 55 (1988); an expanded version is found in *A Neurocomputational Perspective*, pp. 255–79.

land is clearly committed to regarding each of the assertions above—and dozens more on these pages and throughout his article—as false. But surely, our tolerance—and Churchland's tolerance—for contradiction and falsehood has to have a limit somewhere?

DO THE CONTRADICTIONS MATTER?

This objection seems to me both clear and decisive, yet I have learned from experience that many philosophers, not all of them eliminativists, tend to resist it. They would like to allow Churchland to go on using the standard intentional vocabulary in explaining his own theory, without having the resulting contradictions count against him. So I will consider here a few of the ways in which this might be done.

One of the more interesting suggestions begins with the reminder that there are after all well-established ways of talking—say, by Zen practitioners and perhaps by some postmodernists—that involve the deliberate, and deliberately provocative, use of contradiction. If we are wise, we do not immediately write these utterances off as nonsensical; rather, we listen thoughtfully and try to discern the deeper insights that lurk behind the contradictions.

The notion of the eliminativist as Zen master and/or postmodernist is an intriguing one, but it does not fit very well the image the eliminativists themselves seem to want to project. Still, the option remains open: if eliminativists will let us know that they want to have their utterances regarded in this light, we certainly ought to oblige them. It would not do to have the "sound of one hand clapping" drowned out by tedious philosophical chatter!

A more serious suggestion is that my interpretation of the eliminativists' claims is too strong—that in taking them to deny outright that there are such things as beliefs, desires, and the like I have given a forced, extreme reading that creates needless difficulties. One correspondent argued this way: Even though, on Churchland's view, folk psychology is a radically false theory, the folk-psychological concepts can very well survive the demise of the theory and can be used in making true statements. "So it doesn't seem at all obvious that someone must use these common-sense concepts in a way that implies that any particular theory is true. The advocate of EM can say that he's using the common-sense psychological vocabulary in this ordinary sense, akin to the way we speak when we say it's true that the Sun just moved behind that ridge . . . i.e. innocent of false theoretical and ontological implications."[28]

A position such as the one suggested is not only possible; it may actually be held by some philosophers. It's a bit hard to see, though, why it should be called "eliminative materialism": What is it that is being eliminated? In any case, this is clearly not Churchland's own stance. Consider, in this regard, his response to the self-refutation argument: Surely here, if anywhere, it would be incumbent on him to explain that talk about the nonexistence of beliefs, desires, and so on is rhetorical overkill, needing to be reigned in for the purposes of a sober and accurate exposition. Churchland, however, goes in just the opposite direction: "Consider . . . the evident conflict between the eliminativist's apparent *belief* that FP is false, and his concurrent claim that there *are no* beliefs."[29] Here Churchland passes up the golden opportunity to qualify his claim that "there *are no* beliefs"; rather, it's the "*belief* that FP is false" that is merely "apparent"— which is exactly what one would expect on the reading of Churchland given here. And Stephen Stich (who must now be referred to as a "former eliminativist"), lists as the main conclusion of eliminativism that "beliefs, desires, and other posits of folk psychology do not exist."[30]

Another suggestion, perhaps more in line with Churchland's intent though it is not given by him, is that the propositional-attitude terms do not occur in the quotations with their ordinary senses, but rather as stand-ins for the names, as yet unbestowed, of the successor concepts heralded by eliminativism. It is important to realize that *this option is not available.* We simply *have no grasp* of these successor concepts, and cannot use them to make any assertions, no matter how they are named. Indeed, we have no assurance (as Churchland's scenario makes clear) that the roles played by the successor concepts will be even "remotely analogous" to those occupied by the concepts of our present scheme. No. The concepts involved in the quotations from Churchland's article, the only concepts available to him, are precisely the concepts of the commonsense conception renounced by eliminativism. The charge of falsehood and contradic-

28. Don Wacome, in correspondence. I should add, however, that after reading Churchland's "Postscript," Wacome withdrew this "charitable interpretation."

29. Churchland, "Postscript," p. 170.

30. Stephen P. Stich, *Deconstructing the Mind* (New York: Oxford University Press, 1996), p. 4. However, Stich also notes that "some authors use the term *eliminativism* for the view that folk psychology is a seriously mistaken theory," rather than for the ontological thesis that beliefs, desires, etc., do not exist (ibid., p. 83 n. 21).

tion remains.[31] And if a theory which admittedly contains self-contradiction and massive falsehood is not self-refuting, what more does it take?[32]

Churchland, however, does not see it that way. In response to the argument that eliminativism is self-defeating, he asserts that the eliminativist argument against folk psychology is a straightforward *reductio*: "Assume Q (the framework of FP assumptions); argue legitimately from Q and other empirical premises to the conclusion that not-Q; and then conclude not-Q by the principle of *reductio ad absurdum*." He goes on to assert that "if the 'self-defeating' objection were correct in this instance, it would signal a blanket refutation of all formal *reductios*, because they all 'presuppose what they are trying to deny.' "[33] But the situation for Churchland is quite different from that in an ordinary *reductio* argument. In a *reductio* argument, the arguer once having reached the conclusion not-Q *proceeds on the assumption that Q is false*. Indeed, in most cases he never really asserted the truth of Q in the first place, but only presented it as a "provisional assumption." But Churchland, having reached his negative conclusion about folk psychology, nevertheless continues (as we have seen above) to make assertion after assertion *assuming the truth of what eliminativism denies*. There is no parallel to this in the case of the ordinary *reductio* argument; thus Churchland's attempt at self-defense fails.

I suspect some readers will still think I am being uncharitable. "Granted," they may say, "that Churchland is in an awkward position, one in which he can't express himself without apparent self-contradiction. Still, we *are* able to understand, well enough to get on with it, what his eliminative materialism is asserting. And we are also able to understand the locutions contained in the quotations from his article, even though they apparently conflict with what eliminativism asserts. So why can't we just proceed on that basis, and allow the eliminativists the time they say they need to remove the contradictions?"

My reply is that my difficulty lies precisely in understanding the elimi-

---

31. Ironically, Churchland's writings may contain *less* falsehood when judged from the standpoint of the anti-eliminativist than when judged from his own eliminativist standpoint. For the anti-eliminativist, the falsehood lies primarily in the denial of the existence of propositional-attitude states. Once this denial has been given up, the specific assertions concerning propositional attitudes can be judged on their own merits, and many of them may turn out to be true.

32. Unless, of course, we are to be told that contradiction is a problem only from the standpoint of "folk logic," which, like so much else, is destined for replacement in the coming scientific worldview.

33. Churchland, "Postscript," p. 171.

nativist position in such a way that it doesn't immediately self-destruct. I do think I grasp what is meant by "there are no beliefs"—and as we've noted, Churchland clearly is *not* willing to qualify that assertion in such a way that it would not have the radical implications it seems to have. And I find the quotations from his article readily intelligible. What I do *not* find it possible to do, however, is to understand those quotations in some way that does not flatly contradict the basic claims of eliminativism. Those who find themselves able to accomplish this feat are invited to explain how, so the rest of us can both admire and imitate their achievement. Otherwise, the suspicion remains that we are being asked to simply ignore the fact that eliminativists engage in massive self-contradiction.

AUSTERE ELIMINATIVISM AND COGNITIVE PARALYSIS

It may be, however, that somewhere there is an eliminativist who recognizes the untenability of this situation. She realizes that it is just not acceptable to continue writing checks on an account which has supposedly been closed out, in the hope that some day another account may be opened which will give her the resources to pay off all the debts that have accumulated. Nevertheless, she is convinced by the arguments for eliminative materialism, and wants to remain an eliminativist—but she recognizes that in order to do this with integrity she must resolve to live within the conceptual means eliminativism affords her.

What she must do, in order to achieve her goal, is to become an Austere Eliminativist.[34] That is, she will simply *stop using* the idioms of propositional attitudes, assertion, rational acceptability, truth, and other notions eliminativism requires her to renounce. She will content herself with stating, plainly and soberly, those items in her belief-structure (that is, in what a non-eliminativist would identify as her belief-structure) that can be expressed without involving any of the objectionable notions. She will continue, to be sure, to be able to "understand" the locutions of the practitioners of folk psychology: that is to say, she will understand them to the extent one can understand a way of speaking that one has been socialized into but which one has come to regard as irremediably confused. But if pressed with objections drawn from folk psychology, she may find that no

34. I borrow the term from Devitt, but give it a slightly different meaning than his (see Devitt, "Transcendentalism about Content," p. 251).

answer is possible within the limits of Austere Eliminativism; in such cases she will simply remain silent.[35]

Interestingly, the Austere Eliminativist is in a position to benefit from an argument of Churchland's which fails completely to support his own position. We've already seen how, in response to the argument that eliminativism is self-defeating, Churchland asserts that the eliminativist argument against folk psychology is a straightforward *reductio*, no more objectionable than other arguments of that species. This fails as a defense of Churchland because, having reached his negative conclusion about folk psychology, he nevertheless continues to make assertion after assertion *assuming the truth of what eliminativism denies*. But it succeeds as a defense of the Austere Eliminativist, for she takes seriously the renunciation of commonsense psychology demanded by eliminativism, and tries to live by it in her cognitive practice.

All the same, the Austere Eliminativist has a hard way to go. It is questionable whether she can even manage a statement of eliminative materialism itself within the resources available to her. Consider Churchland's assertion that "eliminative materialism is the thesis that our commonsense conception of psychological phenomena constitutes a radically false theory." What is a thesis, if not something asserted? And on Churchland's own account, truth and falsehood have to go. And when he says the theory is "defective," he appeals to standards for rational evaluation which, so far as we can tell, have no place in the Austere Eliminativist's lexicon. Still, there may be some way around these difficulties. Or, it may be that it really doesn't matter; perhaps eliminative materialism is a medicine which need be taken only once: after it has been ingested, the disease is cured and the medicine is no longer required.[36]

In any case, the Austere Eliminativist's discourse is severely restricted.

---

35. Here we are reminded of Wittgenstein: "The correct method in philosophy would really be the following: to say nothing except what can be said, i.e. propositions of natural science—i.e. something that has nothing to do with philosophy—and then, whenever someone else wanted to say something metaphysical, to demonstrate to him that he had failed to give a meaning to certain signs in his propositions. Although it would not be satisfying to the other person—he would not have the feeling that we were teaching him philosophy—*this* method would be the only strictly correct one" (*Tractatus* 6.53).

It should be noted, however, that the Proper Tractarian has considerably greater conceptual resources at his disposal than does the Austere Eliminativist.

36. Once again, the parallel with Wittgenstein is obvious.

She cannot say that she and others *assert* statements, or that they *rationally accept* them, or that the statements are *true* or *false*, or anything else involving these notions or their equivalents. Nor, of course, can she use any propositional-attitude terms. I will make no attempt to construct a sample of austere eliminativist discourse; a brief look at almost any philosophical article will reveal how frequently the proscribed locutions tend to appear, and how much one would lose by renouncing them. Even with a great deal of ingenuity, the Austere Eliminativist's efforts to participate in philosophical discussion are likely to be gravely hindered. And if she similarly restricts her interior monologue (as it seems in good conscience she must), her overall cognitive functioning will undoubtedly be seriously impaired. Baker's term "cognitive suicide," while colorful, is a bit too extreme for such a case: suicide brings about a cessation of functioning which is total and irreversible, neither of which need be the case here. Let us say, then, that the Austere Eliminativist is suffering from *partial, self-imposed cognitive paralysis*.

ELIMINATIVE MATERIALISM: A DIAGNOSIS

John Searle has remarked, "If you are tempted to functionalism, I believe you do not need refutation, you need help."[37] I do not know whether it will be possible for anything I say here to help persons tempted to eliminativism. But a diagnosis is sometimes the first step toward a cure, so I offer in closing a diagnosis of eliminative materialism.

The diagnosis begins by pointing out a parallel with another radical movement in twentieth-century philosophy, namely logical positivism. The early positivists, committed to a rather extreme and dogmatic variety of empiricism, saw clearly that their empiricism simply could not account for large ranges of human thought and discourse—morality, religion, metaphysics, and even (as it turned out) a great deal of scientific discourse. Rather than revise their empiricism to accommodate these data of human experience, they simply ruled the data out of court while leaving the door open for their recovery, as "emotive language," in a form that voided most of their original significance. Eventually, of course, reality reasserted itself. Nowadays ethics, religion, metaphysics, and science all go

37. Searle, *Rediscovery of the Mind*, p. 9.

about their business largely untroubled by the positivist assault, which is well on its way to becoming a distant memory.

My suggestion is that something quite similar to this is going on in the case of eliminative materialism. The commitment here is not so much to empiricism as to a metaphysically realist materialism, which has as a corollary that whatever is real in the world has to be explainable in terms of materialistic science.[38] The task of carrying out this assignment with respect to the phenomena of the mind has been the preoccupation of the philosophy of mind for the past several decades. The enterprise continues to flourish, but there are signs on many fronts of intractable difficulties: consciousness, qualia, mental content, and intentionality, to name a few. There are still optimists who expect materialism to be able to overcome thse difficulties. But if one is inclined to pessimism, the eliminativist strategy begins to look quite attractive. Just as the positivists rejected the data of ethics—or rather, reinterpreted the data in a form which left them barely recognizable—so the eliminativists reject the data embodied in our ordinary experience of the life of the mind. Rather than endure the unresolved tension between theory and data, they boldly proclaim that the data in question are themselves contaminated with bad theory and are thus ineligible for incorporation in a proper science of the mind.

I do not wish to claim that these considerations constitute the entire motivation for the eliminativist project. It is clear that Paul and Patricia Churchland, for example, are involved in a continuing, dynamic, interdisciplinary program of cognitive science research. This research has generated both excitement and significant results, and promises to produce more of both in the future. But eliminative materialism is hardly a necessary condition for cognitive science research, and in fact researchers in the field occupy a fairly wide spectrum of positions in the philosophy of mind. My suggestion is that eliminative materialism in the narrower sense, as defined by Paul Churchland in his 1981 essay, is indeed motivated in the way described.

I believe the diagnosis suggested here can throw light on a couple of situations that might well puzzle anyone first coming to study this field of philosophy. The first puzzle is this: Why do eliminativists insist on such strong and highly counterintuitive claims—for instance, that there are no beliefs, desires, or intentions—when more modest claims would serve

---

38. As we have noted, this is threatened with compromise by Churchland's venture into pragmatism.

many of their purposes equally well? Often when reading eliminativists one receives the impression that their main concern is that brain research may never find anything at all closely corresponding to the desires, beliefs, and so on of folk psychology. Clearly, the empirical researchers in this field must have the freedom to discover what is (or isn't) there to be found, rather than being constrained to find "sentences of mentalese" and other brain correlates of folk-psychological concepts. But why don't the eliminativists content themselves with the observation just made—that brain research may never discover anything closely corresponding to the concepts of folk psychology—rather than making those shocking, and counterintuitive, statements we have come to know as "eliminative materialism"? Is the sheer love of provocation the *only* explanation for this phenomenon?

The second puzzle is this: Why are so many non-eliminativists strongly resistant to the idea that eliminativism has been conclusively refuted? Eliminativism is not, in fact, a very popular view; the number of self-professed eliminativists seems to be quite limited. But there is a considerably larger number of philosophers who, not eliminativists themselves, nevertheless find it important to answer arguments against eliminativism and to insist that the position has not been conclusively refuted. Now of course, some of the arguments against eliminativism *are* bad ones; as we've seen, some of the self-refutation arguments really do beg the question. But even leaving those arguments aside, there are enough genuine difficulties, unfulfilled tasks, unpaid promissory notes, and outrageously implausible claims charged to the account of eliminativism that one might well wonder what keeps the beast alive. Many other views in philosophy, especially the less fashionable ones, are declared dead on the basis of objections that are far less cogent. Does the failure to render this verdict on eliminativism reflect simply an upwelling of fair-mindedness on the part of non-eliminativists, or is there something else at work here?

In order to see how my proposal resolves these puzzles, it will be helpful to have before us an "Argument from Physicalism" formulated by Lynne Rudder Baker. It goes as follows:

(1) Either physicalistic psychology will vindicate . . . the common-sense conception of the mental, or the common-sense conception is radically mistaken.
(2) Physicalistic psychology will fail to vindicate . . . the common-sense conception of the mental.

(3) The common-sense conception of the mental is radically mistaken.[39]

The argument is clearly valid; like all valid arguments, it can be read either as *modus ponens* or as *modus tollens*, the latter being Baker's preference.

Now, consider Churchland's stance in relation to this argument. He has come to believe, as a result of a complex set of considerations, that (2) is likely to be true. Since he takes (1) as axiomatic, this means that (3) is also quite likely to be true. (He never claims more than this.) Suppose, on the other hand, he were to accept that folk psychology is "in principle irreplaceable," that it *could not* be rational to reject beliefs, desires, and the like. This, in Churchland's view, would impose unacceptable constraints on research in cognitive science. The cognitive scientist would then be committed to continue indefinitely the search for the "physicalistically correct" way of naturalizing folk psychology. Rather than cope with this uncongenial and very possibly fruitless task, Churchland affirms the strong eliminativism we have come to know and love/hate. It is not, of course, essential to his project that (3) *is* true; what is essential is that it *may* be true, that it is perfectly acceptable to reject folk psychology should the scientific research come out in support of (2).

Now, what about the non-eliminativists who are so energetic in their defense of eliminativism? Unlike Churchland, they are optimistic about the possibilities of accommodating our commonsense conception of the mental within an acceptably physicalistic psychology. So their considered view is that in all probability (2) and (3) are both false. They are quite open, furthermore, to admitting that there are serious objections to (3), that it is highly implausible, and the like. But they will not accept that (3) has been *conclusively* refuted, and the suggestion that (3) is *self*-refuting draws a particularly strong response. Why is this so? Is there more involved here than a simple assessment of the strength of the various arguments?

I think there *is* more here than meets the eye.[40] Keep in mind that even

39. Baker, *Saving Belief*, p. 6. The ellipses replace parenthetical expressions indicating that the kind of "vindication" that is required is to be specified later; this issue need not concern us at present.

40. Though not, perhaps, in every case. I am happy to accept Bertolet's "boring, but accurate" explanation for his own contribution: "I taught the book and so thought hard about it, thought I saw an important gap in the argument that others

though they are relatively optimistic, these philosophers realize that the evidence to date supplies considerable, though not decisive, support for (2). Now, if (3) had been *conclusively* refuted, it would follow that *if there is a serious possibility that (2) is true, there is an equally serious possibility that (1) is false*. But (1), for many of these thinkers, has a status approximating that of a theological dogma—perhaps a little lower than that of infallible papal pronouncements, but not by much. So in order to avoid the possibility of an article of faith being called into question, it is imperative that (3) be kept alive as an option, however remote the likelihood of its actually being true. So for different reasons, these non-eliminativists reach essentially the same conclusion as Churchland: (3) may in fact be false, but the possibility of its being true needs to be kept alive in case (2) turns out true after all.[41]

This concludes my diagnosis, and it also points the way toward subsequent chapters of the book. For me, as for Baker, the Argument from Physicalism is a *modus tollens*, and evidence supporting (2) is as such evidence against (1). No doubt many will still continue to read the argument in the opposite direction. But if they maintain that the truth of (3) is a serious possibility, I believe they have a huge amount of work to do beyond what has been done thus far. And if, true to their own convictions, they become Austere Eliminativists, they have precious few resources with which to get the job done.

---

seemed to have noticed but didn't explain at all clearly, and so thought I had something worthwhile to say about it" (private communication).

41. So for Churchland, the Argument from Physicalism is a straightforward *modus ponens*. For the non-eliminativists, it is a *modus tollens*, but only the second premise is considered open to falsification. If on the other hand the second premise should prove to be true, they would reluctantly reverse course and run the argument as a *modus ponens*.

The line of thought in the last two paragraphs is indebted to a suggestion from Victor Reppert.

CHAPTER TWO

The Limits of Identity

The physical facts determine all the facts. When everything physical has been explained, everything has been explained, period. Furthermore, all explanations must in the end be given in physical terms; nonphysical explanatory principles are unacceptable. Our aim must be to incorporate all of our knowledge, including and especially our knowledge concerning human beings, into the growing scientific—which is to say, physical—worldview.

Such are the aspirations of contemporary materialism or physicalism (I use the expressions interchangeably). But fulfilling these aspirations has proved to be more difficult than many anticipated. Eliminative materialism is a particularly bold strategy for fulfilling them, but one that (as was argued in the first chapter) faces formidable, perhaps insuperable, objections. On the other hand, eliminative materialism itself arose largely because of the obstacles facing other, less radical, materialist strategies. Mainstream philosophers of mind are agreed that the correct view must consist in some form of physicalism, but sometimes it seems they agree on little else.

There is much in this situation that a nonmaterialist can enjoy and appreciate. In fact, there is some temptation merely to sit back and allow the proponents of different versions of materialism to refute each other. This, however, would not be a promising or constructive strategy. It is unpromising, because however many versions of materialism are refuted and/or abandoned, others will continue to appear to take their place. And it is unconstructive because, unless we are contented skeptics, we want to know what is *true* about the world; it is unsatisfying merely to conclude that certain widely held views are false.

The overall strategy of this book is to present certain objections which, it will be argued, are fatal to any possible form of physicalism, and then to develop a constructive alternative. In view of this strategy, the differences between versions of physicalism are less important than they might be in another context. Still, they can't be wholly ignored, for at least a couple of reasons. For one thing, it will be difficult to be sure that the objections really are effective against all varieties of materialism without at least briefly reviewing the main forms and identifying their central tenets. It will be helpful, furthermore, to be able to identify a "best version" of materialism, one that, among the available candidates, does the best job possible of fulfilling the materialist's aspirations while remaining internally coherent and consistent with known facts. This will then be the version we have primarily in mind in the antimaterialist arguments of subsequent chapters. It may not, in the end, be terribly crucial for the reader to be in agreement as to what this best version of materialism should be; nevertheless the exercise of arriving at such a candidate should prove useful. Our aim, then, is twofold: to identify a version of physicalism which is strong enough in its claims to be accepted by most materialists as a legitimate candidate, and yet not so strong as to be subject to obvious objections on either factual or conceptual grounds.

QUASI-ELIMINATIVIST STRATEGIES: ANALYTIC BEHAVIORISM
AND FUNCTIONALISM

Although eliminative materialism per se is one of the later arrivals on the scene, it becomes evident in retrospect that some earlier (and more widely accepted) versions of materialism share in the tendency to eliminate the mental *in the sense in which we all naturally understand it.* Thus, analytic behaviorism, which holds that mental states, properties, and events can be defined in terms of actual and potential overt behavior, has nothing to say about subjective, conscious experience; we could all be zombies, and it would make no difference to the behaviorist. Analytic behaviorism has been generally abandoned; it is pretty well agreed on all sides that the definitions it calls for can't be produced. This of course need not (and must not) be taken to imply a rejection of the behaviorists' sound insight that our attribution of mental states and properties is closely tied to observable behavior.

If analytic behaviorism is defunct, the same cannot be said of functionalism or of versions of the mind-body identity theory that initially identify mental states with causal-functional states. The key idea here is that mental states are defined in terms of the causes that produce them and the effects they produce: these causes and effects include sensory stimuli as causes and overt behavior as effects, but also internal states of the organism as both causes and effects of the states in question. (This marks a major improvement over behaviorism, because internal states are more directly connected with the "mental" states than are the behavioral states featured in behaviorist definitions; furthermore, the internal states are present in cases where observable behavior is not.) These causal-functional definitions do not by themselves entail materialism, but functionalism supports materialism through a two-stage process: First, as a matter of definition, mental states are equated with causal-functional states. Second, it is proposed that scientific investigation will reveal that the state-types which fill the functional roles are physical.

Oceans of ink have been spilled in the debate over functionalism, and there is little prospect that anything said here can add much to the accumulated wisdom on the subject. We have already noted John Searle's dictum, "If you are tempted to functionalism, I believe you do not need refutation, you need help."[1] For a more moderate response, we may consider Jaegwon Kim's remark that "qualia are intrinsic properties if anything is, and to functionalize them is to eliminate them as intrinsic properties."[2] Eminent authorities could of course be quoted on the other side. As a very minor contribution to the discussion, I pose the following problem: Consider the mental state (one that each of us has been in at one time or another) of being embarrassed about one's appearance. Let's suppose, for the sake of the argument, that the functionalist can successfully pick out this state by a causal-functional description. My question is this: Is the state picked out by the functionalist such that, in standard cases, there is a qualitative "feel" to it—that there is "something it is like to be in" such a state? If not, then the reply seems obvious: a kind of state that does not, in ordinary cases, possess that certain qualitative "feel" cannot possibly be

---

1. John Searle, *The Rediscovery of the Mind* (Cambridge: MIT Press, 1992), p. 9.
2. Jaegwon Kim, "Making Sense of Emergence," *Philosophical Studies* (forthcoming).

the state of being embarrassed about one's appearance. It seems possible, to be sure, that one might be embarrassed about one's appearance but be unaware of that fact, having either suppressed the awareness or been distracted by other stimuli so as not to attend to it. But that I might be embarrassed about my appearance without there being any feeling of embarrassment involved, or any tendency or potential whatever for the embarrassment to be felt in that characteristic way, is something I at least find quite unintelligible.[3] So we would have in this case another instance of a ploy that is, unfortunately, all too common in the philosophy of mind: familiar terminology is retained, but it is redefined in such a way that we are no longer talking about the original subject matter.[4]

Suppose, on the other hand, the answer to my question is Yes. Suppose, that is, that the state that fulfills the causal-functional role of embarrassment over one's appearance also characteristically involves (at least the potential of) *feeling* embarrassed. If so, then the causal-functional state identified by functionalism might very well *be* the state of embarrassment we all experience from time to time. (It certainly is not part of the concept of embarrassment that such a state does *not* fulfill any causal-functional role.) In this case, however, the functionalist doctrine will hardly have fulfilled its intended purpose of naturalizing—or "materializing"—the mental. For it will then be the case that the functionally defined mental states characteristically involve phenomenal properties, but the functionalist theory tells us nothing at all useful about how to understand these properties or how they can be incorporated into a materialistic worldview.[5]

3. Since the feeling of embarrassment might at a given time fade from awareness to the point of being unnoticed it seems plausible that, strictly speaking, it is the *potential* for such a feeling that is essential to being embarrassed, rather than an actual feeling.

4. This seems to be the right place to classify Daniel Dennett's views about these matters. He identifies his theory of content as functionalist: "all attributions of content are founded on an appreciation of the *functional roles* of the items in question in the biological economy of the organism (or the engineering of the robot)" ("Dennett, Daniel C.," in Samuel Guttenplan, ed., *A Companion to the Philosophy of Mind* [London: Blackwell, 1995], p. 239). And he rejects consciousness as a requirement for intentionality, as is shown by the reference to robots in the quotation. Presumably, then, he would hold that a robot, if sufficiently complex, could be embarrassed about its appearance without being in any way conscious or even potentially conscious.

5. This is, of course, a version of the much-discussed phenomenal property or "qualia" objection to functionalism; readers are invited to supplement it with their own favorite versions of this objection.

A maneuver that has recently become somewhat popular needs to be taken account of here. Some philosophers have thought it plausible to accept a functionalist account specifically for *intentional* states, while conceding the "phenomenal properties" objection for qualia and giving some other account (e.g., token-identity) for these states. Once again let us concede, for the sake of the argument, that the functionalist can successfully identify intentional states in causal-functional terms. Even so, there are at least two reasons why this maneuver does not succeed in disposing of the materialist's problem with intentional states. For one thing, some intentional states *do* essentially involve qualitative "feels"; being embarrassed about one's appearance is an example of such a state. In general, emotional states involve *both* a particular qualitative "feel" and *also* an intentional reference to whatever it is that one is happy, sad, proud, indignant, or embarrassed about. For these particular states, then, the "divide and conquer" strategy can't succeed.

An even more fundamental problem, however, is encountered in the "aboutness" which is the very essence of an intentional state. There just is such a thing as thinking about something, worrying or hoping that something may happen, believing that so-and-so is the case, deciding on a certain course of action—and in such cases one ordinarily has a distinct, conscious awareness of the "intentional object" of one's mental state.[6] This sort of mental state is utterly familiar to each one of us, and if anyone were to claim not to understand what is meant, such a claim would be entirely lacking in credibility. Furthermore, the claim that a person is in such an intentional state is clearly *not equivalent*, logically or conceptually, to any causal-functional description of the person. The causal-functional properties of the state identified by functionalism can be *completely described and explained* in terms of the physical structure and behavior of the state in question and its relations to other physical states. There simply is no place in such a description for the "aboutness" which is essential to intentional states as such. And so we can pose a question parallel to the one asked previously: Are these causal-functional states such that they essentially involve "aboutness"? If they do not involve "aboutness," then they just are not intentional states—once again,

---

6. This is not to deny, of course, that one may sometimes be in an intentional state and be unaware of the fact. But these "nonconscious intentional states" are identified as such only because of their similarities to the more typical cases, in which one *is* aware of the intentional object.

the subject has been changed.[7] If they do, then a crucial, and logically essential, aspect of these states has been left unexplained; we still have no idea how the "aboutness" is to be incorporated into a materialistic worldview.

In sum: either the causal-functional states identified by functionalism characteristically involve conscious "feels" (in the case of qualia) and "aboutness" (in the case of intentional states) or they do not. If they do not, then functionalism is not a theory about the mental at all; the name remains, but the subject has been changed. If they do, then functionalism may indeed be a theory (possibly even a true theory) about the mental, but it does next to nothing to explain two of the most salient features of the mental states in question, namely subjective consciousness and intentional reference. In this case, there seems to be a need for a theory that more directly addresses the key issue concerning the relationship between consciousness and its physical embodiment.

MIND-BODY IDENTITY THEORIES

The leading candidate for such a theory is probably the mind-body identity theory in its various guises. Identity theories have been prominent since the 1950s and still remain active contenders, especially in the form of token-identity theories. One strength of such theories is that they begin by frankly acknowledging the reality of mental phenomena, instead of denying them, ignoring them, or redefining them out of existence. These phenomena are then said to be identical with some undeniably physical phenomena, thus incorporating the mental into the physicalistic worldview.

7. Searle's remarks on this deserve to be quoted at length. He states,

Any attempt to reduce intentionality to something nonmental will always fail because it leaves out intentionality. Suppose for example that you had a perfect causal externalist account of the belief that water is wet. This account is given by stating a set of causal relations in which a system stands to water and to wetness and these relations are entirely specified without any mental component. The problem is obvious: a system could have all of these relations and still not believe that water is wet. This is just an extension of the Chinese room argument, but the moral it points to is general: You cannot reduce intentional content (or pains or "qualia") to something else, because if you could they would be something else, and they are not something else. The opposite of my view is stated very succinctly by Fodor: "If aboutness is real, it must really be something else." . . . On the contrary, aboutness (i.e., intentionality) is real, and it is not something else (*Rediscovery of the Mind*, p. 51).

An important move made early on in this process was the shift from speaking of mental *objects* to mental *events, processes,* and *states.* If one thinks in terms of mental objects, it is possible to reason thus: "When I am seeing some round, orange object, there is in my mind something actually round and orange—call it a sensation, a sense-datum, or what you will. But when my brain is examined, there is nothing round and orange to be found there. So since my mind contains this round, orange object and my brain doesn't, the two can't possibly be identical." J. J. C. Smart saw the way around this objection: rather than recognize sensations or sense-data as mental *objects,* we should speak of the mental *event* or *process* of sensing in a certain way. Smart put it like this: "If it is objected that the after-image is yellowy-orange, my reply is that it is the experience of seeing yellowy-orange that is being described, and this experience is not a yellowy-orange something. So to say that a brain process cannot in fact be yellowy-orange is not to say that a brain process cannot in fact be the experience of having a yellowy-orange after-image. There is, in a sense, no such thing as an after-image or a sense-datum, though there is such a thing as the experience of having an image."[8] To be sure, the materialist is not yet home free; there remains the task of supporting the identity between the experience and the brain process. But whatever difficulties may lie in wait (and as we will see, they are formidable), they are preferable to looking for that round, orange thing in the brain.

So mental events, states, and processes are to be identical with physical events, states, and processes. And this is most naturally taken to mean that *kinds* or *types* of mental states are identical with kinds and types of physical processes—thus, "type-identity" theories. The stereotyped example of such an identification is: pain = C-fibers firing; for a more up-to-date illustration, we might take Frances Crick's suggestion: visual attention = correlated firing of neurons in the visual cortex.[9] Are such proposals credible? They encounter an immediate obstacle in what I shall call the Descartes Objection.[10] It seemed to Descartes (and it seems to many of us still) that he could conceive of himself existing—and, we may add, directing his visual attention here or there—without conceiving of anything physical existing at all. A fortiori, he could conceive of this without con-

8. J. J. C. Smart, "Sensations and Brain-Processes," in John O'Connor, ed., *Modern Materialism: Readings on Mind-Body Identity* (New York: Harcourt, Brace, and World, 1969), pp. 32–47; the quotation is from p. 42.

9. See Frances Crick, *The Astonishing Hypothesis: The Scientific Search for the Soul* (New York: Scribner's, 1994), p. 311.

10. See *Meditations* 2 and 6.

ceiving of the correlated firing of neurons in his visual cortex, of which he had never so much as heard. And this, it may seem, is sufficient to show that the property of visual attention is distinct from the property of having neurons firing in a correlated fashion in one's visual cortex.[11] Descartes himself, of course, thought he could in this way prove directly the substantial distinction between mind and body. For reasons that will be explained in Chapter 5, I don't think this works. But the Descartes Objection does seem to pose a serious obstacle to the identification of mental and physical properties of persons.

Contemporary identity theorists are not unaware of the Descartes Objection; they meet it by distinguishing between properties and predicates, or properties and concepts. Clearly the *concept* of visual attention is distinct from the *concept* of the correlated firing of neurons, but it still may be (so it is claimed) that the *properties* involved are identical. Which leads, of course, to the question: What is the criterion for the identity of properties?

A useful approach to this question is offered by Philip Bricker,[12] who proposes three different conceptions of properties: the extensional conception, the intensional conception, and the structured conception. On the extensional conception, properties are identical if and only if they have the same extension; a property is identical with the class of its instances. So if in fact all and only those animals have hearts that have kidneys, it follows that *being a creature with a heart* and *being a creature with kidneys* are identical properties. On the intensional conception, properties are identical if and only if they are *necessarily* coextensive. On this conception, *being a creature with a heart* and *being a creature with kidneys* are *not* identical properties, since even if it is in fact true that all creatures with hearts have kidneys, and vice versa, it is clearly metaphysically possible that there be an exception to this rule. Finally, the structured conception of properties is explained by Bricker as follows:

> A third response is to hold that the properties expressed by 'is a creature with a heart' and 'is a creature with kidneys' are different because they are structured entities with different constituents: the property expressed by 'is

11. For an excellent recent presentation of the Descartes Objection, see Charles Taliaferro, *Consciousness and the Mind of God* (Cambridge: Cambridge University Press, 1994), pp. 48–64.

12. Philip Bricker, "Properties," in Donald M. Borchert, ed., *The Encyclopedia of Philosophy Supplement* (New York: Macmillan, 1996), pp. 469–73.

a creature with a heart' has the property expressed by 'is a heart' as a constituent; the property expressed by 'is a creature with kidneys' does not. On this response properties have a quasi-syntactic structure that parallels the structure of predicates that express them. Call two predicates isomorphic if they have the same syntactic structure and corresponding syntactic components are assigned the same semantic values. On a structured conception of properties, properties expressed by isomorphic predicates are identical; properties expressed by nonisomorphic predicates are distinct.[13]

One might naturally ask, Which of these is the *correct* conception of properties? Bricker's response is that we need not and should not choose; rather, we should apply whichever conception is appropriate to the particular context under discussion.

Our present context, of course, is that of mind-body type-identity theories. On the structured conception of properties, such theories fail immediately. 'Is attending visually to something' and 'has correlated firings of neurons in the visual cortex' are in no way isomorphic predicates, and the failure of isomorphism will no doubt become even more striking as more of the physiological detail is filled in. For this conception, the Descartes Objection is decisive.

On the extensional conception, type-identity theories have a better chance, but it may be objected that the conception of property-identity here is just too weak to be plausible. It simply is not credible, in terms of our preanalytical notion of what a property is, that the properties *being a creature with a heart* and *being a creature with kidneys* are identical.[14] And if we hold (as I think we must) to the view that identity-statements must be necessary, then the relation between these properties cannot be one of identity, since (as noted above) it is clearly metaphysically possible that there be a creature with a heart but no kidneys, or vice versa. But on the extensional conception, the properties are "identical" or not depending on whether such an exception actually exists, thus making their identity a contingent matter.

The intensional conception of properties, on the other hand, is more promising as an account for use with type-identity theories. To be sure,

13. Ibid., p. 470.

14. Bricker gives no examples of ordinary discourse about properties that is naturally interpreted in terms of the extensional conception; he does state that such a conception is "adequate to the semantic analysis of mathematical language and extensional languages generally" (ibid.).

there is support from ordinary usage for the view that *being a polygon with three sides* and *being a polygon with three angles* are distinct properties, as one would hold on the basis of the structured conception of properties. But a case can also be made for saying that what we have here is the same property described in two different ways. Could mental and physical properties be like this? They couldn't be *exactly* like this: *being a polygon with three sides* and *being a polygon with three angles* are logically deducible each from the other, which clearly is not the case for mental and physical properties. The mind-body property-identities must rather be understood as *theoretical identities,* of the same kind as Water = H_2O. Our ordinary *concept* of water is clearly distinct from our *concept* of H_2O; nevertheless, the stuff we call "water" *is in fact* H_2O, and (arguably) is not distinct from H_2O in any possible world. If mental and physical properties are said to be identical in *this* sense, the Descartes Objection is no longer decisive.

If mind-body type-identity is spelled out in terms of theoretical identity, it seems a conceivable view—but is there any reason to think it is *true*? What sort of evidence could we have that would make this view more than a theoretical possibility? There could, of course, be evidence of *correlations* between mental states and brain-states; in fact, some of these correlations are already beginning to be established. But such correlations are of little use in distinguishing between mind-body theories; even Cartesian dualists can welcome them. It would be extremely helpful, on the other hand, if it were possible to *deduce* the properties of mental states from the properties of their physical correlates; this would generate serious support for the identity theory, comparable to the evidence which convinces everyone that water is, in fact, H_2O.[15] But such deductions not only are not possible at present; there doesn't seem to be any prospect of their becoming possible, even if we were able to map out a fairly complete account of mind-brain correlations. Indeed, this claim may still be too weak: it may be that it is simply inconceivable for mental predicates to be deduced from a theory that is "physical" in anything approximating our present understanding of that notion. And failing this, the type-identity theory seems at best a highly speculative metaphysical possibility.

The widespread abandonment of type-identity theories has not, however, been due primarily to objections such as those stated above. The

15. The importance of this procedure in cases of reduction is emphasized by Kim in "Making Sense of Emergence," pt. 3. For additional discussion, see the next section of this chapter.

main reason for the decline of these theories is to be found rather in the problem of the "multiple realizability" of mental states. As Hilary Putnam explains, the type-identity theorist

> has to specify a physical-chemical state such that *any* organism (not just a mammal) is in pain if and only if (a) it possesses a brain of a suitable physical-chemical structure; and (b) its brain is in that physical-chemical state. This means that the physical-chemical state in question must be a possible state of a mammalian brain, a reptilian brain, a mollusc's brain (octopuses are mollusca, and certainly feel pain), etc. At the same time, it must *not* be a possible (physically possible) state of the brain of any physically possible creature that cannot feel pain. Even if such a state can be found, it must be nomologically certain that it will also be a state of the brain of any extraterrestrial life that may be found that will be capable of feeling pain before we can even entertain the supposition that it may *be* pain.[16]

The point, surely, is clear enough. We know of some cases, and can readily imagine others, in which there are creatures capable of psychological states that are relevantly similar to those experienced by humans but whose physiology differs markedly from the human. The supposition that "parallel evolution, all over the universe, might *always* lead to *one and the same* physical 'correlate' of pain" is, as Putnam states, an "ambitious hypothesis."[17]

It may be, though, that Putnam is still understating the difficulty. He seems to be thinking of other sentient organisms *in our universe,* or at least in a universe that shares its physical laws with ours. But why suppose that all possible universes are like that? May there not be possible worlds sharing *none* of the physical laws, and none of the same physical elements, with our universe, in which nevertheless sentience exists? To speculate even more wildly, why couldn't there be a universe so drastically unlike ours that in it mental states are enjoyed by Cartesian souls? This supposition may be unwelcome to materialists, but they need better reason than that to declare it metaphysically impossible. Putting all these considerations together, it seems the theory that mental events are type-identical

16. Hilary Putnam, "The Nature of Mental States," in David M. Rosenthal, ed., *Materialism and the Mind-Body Problem* (Englewood Cliffs, N.J.: Prentice-Hall, 1971), p. 157.

17. Ibid., p. 158.

with physical events just can't be sustained; the objections outweigh by far the reasons some have for wishing it were true.

The demise of type-identity theories need not lead to the abandonment of mind-brain identity; there are still token-identity theories to be considered. Token-identity theories, indeed, seem ideally suited to meet the challenge of multiple realization. It may be that not all instances of pain, or of visual attention, can be identified with instances of some one physiological property. But may it not still be the case, that *each particular* instance of pain is identical with some physical event in the suffering organism? The general form of this is: mental event-tokens are identical with physical event-tokens. But this, of course, raises the question of identity conditions for events.

One well-known answer to this question comes from Donald Davidson; it states that an event is identical with a (differently described) event if and only if their causes and effects are identical. To state it formally,

$$(x = y) \text{ if and only if } ((z) \ (z \text{ caused } x \leftrightarrow z \text{ caused } y) \text{ and } (z)(x \text{ caused } z \leftrightarrow y \text{ caused } z)).[18]$$

This criterion readily facilitates the adoption of a token-identity theory, and may have been crafted partly for that purpose. There has been extensive debate about Davidson's view, and about the theory of events generally, going far beyond the scope of the present discussion. But one difficulty has surfaced which seems to show conclusively that Davidson's theory as stated is unsatisfactory. Put briefly, the theory is circular: causal relations are identified by the events they relate, and events are identified by the causal relations in which they stand. Clearly, one or the other has to be identified in some other way, so as to enable us to break into this closed circle. And since there seems to be no prospect of identifying causal relations except as relations between events, it seems that it will have to be the events themselves which are characterized in some other way.[19]

Such an alternative is supplied by Jaegwon Kim's theory of events. Ac-

18. Donald Davidson, "The Individuation of Events," in *Essays on Actions and Events* (Oxford: Clarendon, 1980), p. 179.

19. Davidson himself eventually abandoned his own proposal for this reason. See his "Reply to Quine on Events," in E. Lepore and B. McLaughlin, eds., *Actions and Events: Perspectives on the Philosophy of Donald Davidson* (Oxford: Basil Blackwell, 1985), pp. 172–76.

cording to Kim's account, we should think of both events and states[20] as *exemplifications by substances of properties at a time.*[21] More formally, we have:

> *Existence condition*: Event [x, P, t] exists just in case substance x has property P at time t.
>
> *Identity condition*: [x, P, t] = [y, Q, t'] just in case x = y, P = Q, and t = t'.[22]

(The account could easily be extended to include relational events.) This account does not suffer from the kind of circularity that we have seen in Davidson's. It has, however, been subjected to criticism on various grounds, one of which is that it makes events too numerous. Davidson has pointed out that, on Kim's account, no "stabbing is ever a killing, or the signing of a cheque the paying of a bill."[23] This multiplication of events can, however, be alleviated to some extent if we avail ourselves of a suggestion made by Richard Swinburne. Swinburne contrasts "intrinsic descriptions" of events—descriptions "as what they are in themselves, quite apart from their causes, effects, or what they are in view of the circumstances of their occurrence,"[24] from extrinsic descriptions in terms of their causes, effects, and circumstances. Thus, "Brutus stabbing Caesar" gives an intrinsic description; "Brutus killing Caesar" describes the same event in terms of its effects. Given this distinction, we can say that "two extrinsic descriptions pick out the same event if they both pick out an event with the same intrinsic description. An extrinsic description picks out an event with a certain intrinsic description if the latter event satisfies

---

20. "Events" as Kim explains them include persisting *states* as well as *changes*. The definition could presumably be modified to require that events should be changes, though this would involve the nontrivial task of distinguishing "real" from "Cambridge" changes.

21. Jaegwon Kim, "Events as Property Exemplifications," in *Supervenience and Mind: Selected Philosophical Essays* (Cambridge: Cambridge University Press, 1991), p. 34.

22. Ibid., p. 35.

23. Davidson, "Reply to Quine on Events," p. 170.

24. Richard Swinburne, *The Evolution of the Soul* (Oxford: Clarendon, 1986), p. 52f.

the former description in virtue of its causes, effects, and circumstances of occurrence."[25]

The multiplication of events can be reined in still further by another suggestion of Swinburne's: "One description may be fuller than another in describing the property in more intrinsic detail. S_1 being P_1 at t_1 is the same event as S_2 being P_2 at t_2 where both descriptions are intrinsic descriptions even if $P_1 \neq P_2$ so long as P_2 is a determinate of the determinable P_1 (i.e. it is a particular way of being P_1) and where $S_1 = S_2$ and $t_1 = t_2$. Thus, if I move slowly, my moving at noon is the same event as my moving slowly at noon."[26] This seems an intuitively sound way of assessing the identity of events, and it goes some way toward curbing the excessive multiplication of events that seems to result from Kim's theory.

These maneuvers do not, however, limit the multiplication of events enough to save the token-identity theory. As Davidson observed, Kim "must also hold that if psychological predicates have no coextensive physical predicates [e.g., if multiple realization is possible], then no psychological event is identical with a physical event."[27] Even with the modifications suggested by Swinburne, it will be impossible to identify mental events with their corresponding brain-events. And since Kim's theory of events (and variants thereof) is the most widely accepted account that is currently available, prospects for even a token-identity theory begin to look dim.

THE SUPERVENIENCE OF MIND ON BODY

In recent years, philosophers who wish to be physicalists but are dubious about the prospects of the identity theory have often turned to the notion that mental properties are supervenient on physical properties. The notion of supervenience is usually credited to the moral philosophy of G. E. Moore, who held that (1) moral properties are not reducible to nonmoral properties (as is shown by the refutation of ethical naturalism), but that (2) no two situations that are identical in all nonmoral properties can differ in a moral property. Thus, the moral properties of situations "supervene" on their nonmoral properties. (The origin of this use of the term "supervenience" is traced back to R. M. Hare.)

The core idea of supervenience is thus that of one set of properties de-

25. Ibid., p. 53.
26. Ibid.
27. Davidson, "Reply to Quine on Events," p. 170.

termined by and dependent on another, a notion that is readily adapted to the physicalist conception of the mind-body relationship. Indeed, it may seem that supervenience represents some sort of irreducible minimum for physicalism; a theory which allows mental properties to vary independently of any physical basis could not be physicalist or materialist in any meaningful sense.[28] But while the general idea may seem clear and compelling, the precise formulation of the supervenience relation involves certain difficulties. We shall begin by addressing this problem of formulation and then go on to consider several additional questions.

The Meaning of Supervenience

What exactly is the right way to understand mind-body supervenience? A number of philosophers have been involved in trying to clarify this question. Undoubtedly the leader in this effort has been Jaegwon Kim, and we will proceed by reviewing some of his suggestions. We begin with the notion of "weak supervenience," a view which Kim finds in the writings of R. M. Hare and Donald Davidson. Kim formulates weak supervenience as follows:

> A *weakly supervenes* on B if and only if necessarily for any x and y if x and y share all properties in B then x and y share all properties in A— that is, indiscernibility with respect to B entails indiscernibility with respect to A.

He goes on to say, "We shall call A the supervenient family and B the supervenience base (family); properties in A are supervenient properties, and those in B are the base properties."[29] Kim points out, however, that weak supervenience is *too* weak to sustain some of the intended applications of the notion of supervenience. The reason for this is that weak supervenience holds constant the relation between base properties and supervenient properties *within a given possible world,* but permits the

28. There is admittedly a complication here, in that our ascription of mental properties (e.g., intentional states) is in part determined by contextual factors rather than by the internal states of the organism—thus, "meaning externalism." The mental states which supervene upon physical states must be understood as *internal or intrinsic states* of the organism in question; it is acknowledged that external factors may affect how these states are (correctly) described.

29. Kim, "Concepts of Supervenience," in *Supervenience and Mind,* p. 58.

relation to vary freely across possible worlds.[30] If, for example, the B-properties are traits of character and the A-properties comprise a man's being good or evil, it is possible that "in this world anyone who is courageous, benevolent, and honest is a good man, but in another possible world no such man is good; in fact, every such man is evil in this other world."[31] As applied to the mind-body relationship, weak supervenience allows for some bizarre possibilities; for example, "In a world that is just like this one in all physical details, unicellular organisms are all fully conscious, while no humans or other primates exhibit mentality."[32]

Since this is obviously unsatisfactory, a stronger supervenience relation must be sought, one that holds the relation between base and supervenient properties constant across possible worlds. One relation that does this is "global supervenience," which may be defined as follows:

> Worlds that are indiscernible in respect of subvenient properties are indiscernible in respect of supervenient properties.[33]

While global supervenience does hold the relation between supervenient and subvenient properties constant across possible worlds, the relation remains excessively fragile. Kim explains:

> First, this form of covariance permits worlds that differ minutely in subvenient properties to differ drastically in respect of supervenient properties. Thus, global covariance of the mental with respect to the physical is consistent with there being a world that differs from this world in some insignificant physical detail (say, it contains one more hydrogen atom) but which differs radically in psychological respects (say, it is wholly void of mentality). Second, global covariance as explained fails to imply weak covariance; that is, it can hold where weak covariance fails. This means that

30. What makes this possible is that the variables 'x' and 'y' range only over *actual* individuals; thus they must both exist within a single possible world. The formula thus says nothing about comparison of objects with the same B-properties in different possible worlds.

31. Ibid., p. 60.

32. Kim, "Supervenience as a Philosophical Concept," in *Supervenience and Mind*, p. 143.

33. Ibid., p. 154.

psychophysical global covariance can be true in a world that contains exact physical duplicates with divergent psychological characteristics . . . [34]

Once again, this relation is obviously too weak to do the work required of it. Clearly, some stronger supervenience notion is needed.

In order to formulate this stronger notion, Kim begins by providing another formulation of weak supervenience:

> A *weakly supervenes* on B if and only if necessarily (for any property F in A, if an object x has F, then there exists a property G in B such that x has G), and if any y has G it has F.[35]

This is, as he shows, equivalent to the formulation quoted above. And it has the additional merit that the limitation of the supervenience relation to a given world is more transparent: the final clause says that "if any y [actually] has G it has F," but fails to account for *possible* individuals, as well as actual individuals that *might* have G but in fact do not—in other words, those that possess G in some possible world different from the actual world. And this sets the stage for the stronger supervenience relation we are seeking:

> A *strongly supervenes* on B just in case, necessarily, for each x and each property F in A, if x has F, then there is a property G in B such that x has G, and *necessarily* if any y has G, it has F.[36]

This avoids the weaknesses noted for both weak supervenience and global supervenience, and it appears to present a good candidate for a notion of supervenience suitable to the mind-body relation. There is, however, one aspect of supervenience which may not be captured adequately by the formula given; this is the idea that the supervenient properties *depend on* the base properties. This idea is not clearly expressed by the definition of strong supervenience, and Kim devotes some effort to an attempt to see how it can be expressed appropriately.[37] We won't go into these discussions

34. Ibid., p. 155.

35. Kim, "Concepts of Supervenience," p. 64 (parentheses added for clarity).

36. Ibid., p. 65.

37. See, for example, Kim, "Supervenience as a Philosophical Concept," pp. 142–49, and "Postscripts on Supervenience," in *Supervenience and Mind*, pp. 165–69.

here, but will simply note that dependence is indeed included as part of the idea of supervenience.

One further question about this supervenience relation needs to be resolved. The definition of strong supervenience contains two occurrences of the modal term 'necessarily.' Kim leaves it undecided how this term is to be interpreted; either occurrence may represent either metaphysical necessity, or nomological necessity (or there may be still other possibilities), and we will get difference supervenience relations depending on which concept of necessity is invoked. Kim states, "For psychophysical supervenience it is possible to interpret the first occurrence as metaphysical necessity and the second as nomological necessity; it is also possible to interpret both as metaphysical, or both as nomological."[38]

Now, consider again the first clauses in the definitions of both weak and strong supervenience: "A *weakly (or strongly) supervenes* on B if and only if necessarily for any property F in A, if an object x has F, then there exists a property G in B such that x has G. . . ." If 'necessarily' here refers to metaphysical necessity, this clause states that *in no possible world* does there exist an object having the supervenient property without the base property—in the present case, in no world is there an object with mental properties but no physical properties. But this is suspect when applied to the mind-body relationship, for some of the same reasons discussed in the section on type-identity theories. It may be thought plausible that *in a world such as ours*, there can indeed be no mental properties without physical base properties. But it is extremely difficult to rule out a priori the possibility of a world with different physical laws, in which all of the physical properties are different than those of our world, in which mentality nevertheless exists. It is also difficult to establish the impossibility of a world in which mind exists *but nothing physical exists at all.* (So far as I know, no one has argued convincingly that Berkeley's world is logically impossible.) In view of these possibilities, it seems best to restrict the first clause of the definition to worlds that are *nomologically possible*, worlds that share their laws of nature with our own world.

Should the final clause of the definition be similarly restricted? This

---

In the latter essay Kim argues that there is no one concept of dependence that applies to supervenience relations in general; he expresses hope that the supervenience of mind on body can be shown to be an instance of mereological supervenience.

38. Kim, "Concepts of Supervenience," p. 66.

could certainly be done, but it may not be necessary. If, to be sure, there are possible worlds that have in them the same sorts of physical properties as our own, but in which the laws of nature are different in important ways, then it is plausible that in some such world the physical properties that in our world provide the supervenience base for, say, believing that Bill Clinton is the president of the United States might provide instead the base for some entirely different mental state. But it may be plausible to hold that sameness of physical properties entails sameness of laws: laws, on this view, are expressions of the inherent causal powers of the kinds of objects that exist, so that objects with different causal powers (and, therefore, governed by different laws) would not share their physical properties with the objects of our world. On this perspective (which seems to me a plausible one), the final clause can retain its unrestricted generality: In no possible world is there an object which shares its physical properties with an object of our world, but fails to share that object's mental properties. For present purposes, then, the second occurrence of 'necessarily' will be understood as referring to metaphysical necessity, but readers who find the assumed view of causality implausible should feel free to read it as nomological necessity.

Our final definition of strong supervenience, then, is as follows:

> A *strongly supervenes* on B just in case, *in every nomologically possible world*, for each x and each property F in A, if x has F, then there is a property G in B such that x has G, and *necessarily* if any y has G, it has F.

In subsequent sections we shall explore some additional questions about this supervenience relation.

Is Reduction Possible?

Does strong supervenience allow mental properties to be reduced to physical properties? The notion of reduction is, of course, an extremely contentious one in the context of the mind-body relationship. Many philosophers, especially those motivated by the "unity of science," have taken for granted that reduction is a desirable objective and have sought to show how it can be achieved. Others, however, have seen the reduction of the mental to the physical as a threat to the "humanistic" qualities of human beings, and this motivates resistance to the idea of reduction. Our question here, however, is not whether reduction is desirable but whether it is possible, given the mind-body relation as defined by strong supervenience.

Once again we turn to Jaegwon Kim. Kim notes that supervenience has often been seized upon as a *nonreductive* version of physicalism. In his APA presidential address, "The Myth of Nonreductive Materialism," Kim argues that this interpretation is misguided. He approaches this, in part, by recurring to Putnam's argument for the multiple realizability of mental states. Kim concedes that multiple realizability creates a serious difficulty for type-identity theories. But he points out that, in arguing for multiple realizability, Putnam in effect assumes a strong sort of connection between mental and physical states—a connection that may well suffice to underwrite a claim to reduce the mental to the physical.

> Putnam appears to be assuming this: *a physical state that realizes a mental event is at least nomologically sufficient for it.* . . . This generates laws of the form, "$P_i \rightarrow M$," where M is a mental state and P_i is a physical state that realizes it. Thus, where there is multiple realization, there must be psychophysical laws, each specifying a physical state as nomologically sufficient for the given mental state. Moreover, Putnam's choice of examples . . . which are either biological species or determinate types of physical structures ("extraterrestrials"), and his talk of "species-specificity" and "species-independence" suggest that he is thinking of laws of a somewhat stronger form, "$S_i \rightarrow (M \leftrightarrow P_i)$," which, *relative to species or structure* S_i, specifies a physical state, P_i, as *both necessary and sufficient* for the occurrence of mental state M. A law of this form states that any organism or system, belonging to a certain species, is such that it has the given mental property at a time if and only if it is in a certain specified physical state at that time. We may call laws of this form, "species-specific bridge laws."[39]

Kim goes on to argue that while laws of this sort cannot support a global reduction of psychology, they can support species-specific or local reductions. He claims, furthermore, that such local reductions are the most

---

39. Kim, "The Myth of Nonreductive Materialism," in *Supervenience and Mind*, p. 273. Kim goes on to note that "species" may actually be too broad here, given individual differences in the localization of brain function. (They might also be too narrow in some instances; there may, for example, be no difference in the brain-states in which pain is realized in closely similar species of sparrows.)

common type of reduction in science generally: "The multiple realization argument, if it works, shows that a global reduction is not in the offing; however, local reductions are reduction enough, by any reasonable scientific standards and in their philosophical implications."[40]

It remains to be noted that strong supervenience has exactly the same implications for the existence of species-specific bridge laws as Kim found in Putnam's discussion, and thus it provides equally strong support for the possibility of local reduction of the mental to the physical. Kim concludes that "supervenience is not going to deliver to us a viable form of nonreductive materialism."[41]

Yet, this may not be Kim's last word on the subject. In a more recent paper, "Making Sense of Emergence," Kim sets out to explore the emergentists' contention that mental states are irreducible to physical states. In the process of doing this, he explicates a concept of reduction that differs significantly from the standard conception that is due to Ernest Nagel.[42] Somewhat abbreviated, the model Kim offers for the reduction of property E to base properties B is as follows:

> Step 1: E must be *functionalized*—that is, E must be construed, or reconstrued, as a property defined by its causal/nomic relations to other properties, specifically properties in the reduction base B. . . .
>
> Step 2: Find realizers of E in B. If the reduction, or reductive explanation, of a particular instance of E in a given system is wanted, find the particular realizing property P in virtue of which E is instantiated on this occasion in this system; similarly, for classes of systems belonging to the same species or structure types. . . .
>
> Step 3: Find a theory (at the level of B) that explains how realizers of E perform the causal task that is constitutive of E (i.e., the causal role specified in Step 1). Such a theory may also explain other significant causal/nomic relations in which E plays a role.[43]

Kim illustrates this conception with a discussion of the reduction of genes to DNA:

40. Ibid., p. 275.

41. Ibid., p. 279.

42. See Nagel's *The Structure of Science* (New York: Harcourt, Brace, and World, 1961). Kim credits David Armstrong for the fundamental ideas of the conception he advocates.

43. Kim, "Making Sense of Emergence," sec. 3.

The property of being a gene is the property of having some property (or being a mechanism) that performs a certain causal function, namely that of transmitting phenotypic characteristics from parents to offspring. As it turns out, it is the DNA molecule that fills this causal specification ("causal role"), and we have a theory that explains just how the DNA molecule is able to perform this causal work. When all of this is in, we are entitled to the claim that the gene has been reduced to the DNA molecule.[44]

Kim emphasizes the difference between this conception of reduction and Nagel's conception.[45] On that conception, a crucial issue concerns the existence of appropriate "bridge laws" connecting the properties of the domain to be reduced with those of the base domain. (This issue surfaced in our previous discussion in the form of the "species-specific bridge laws.") From Nagel's point of view, the availability of such bridge laws is crucial in determining the possibility or impossibility of reduction. But from the emergentist viewpoint embodied in Kim's revised conception of reduction,

the bridge laws, far from being the enablers of reduction (as they are in Nagel reductions) are themselves among the targets of reduction. Why is it that pain, not itch or tickle, occurs when a certain neural condition (e.g., C-fiber stimulation) obtains? Why does not pain accompany conditions of a different neural type? . . . These are the kinds of explanatory/reductive questions with which the emergentists were preoccupied. And I think they were right. The "mystery" of consciousness is not dispelled by any reductive procedure that, as in Nagel reduction, takes these bridge laws as brute unexplained primitives.[46]

For reduction in Kim's sense, the crucial question is not the availability of bridge laws but rather the possibility of giving a functional interpretation of the property that is to be reduced—the first step of the process of reduction, to be followed by the hard scientific work of steps 2 and 3. And so the reducibility of mental states will depend on whether or not they can be given such a functional interpretation.[47] At this point Kim is con-

44. Ibid.
45. These two concepts of reductions are discussed, and contrasted, further in Kim's *Mind in a Physical World* (Cambridge: MIT Press, 1998), chap. 4.
46. Ibid.
47. Kim states, "If the functionalist conception of the mental is correct—correct

tent to "affirm, for what it's worth, my own bias toward [irreducibility]: qualia are intrinsic properties if anything is, and to functionalize them is to eliminate them as intrinsic properties."[48]

In view of all this it seems that the reductive or nonreductive character of strong supervenience depends on the concept of reduction that is accepted. If one is content with Nagel-style reduction, in which bridge laws are taken as primitive, then strong supervenience entails reducibility, albeit local rather than global reducibility. If on the other hand one insists on the stronger conception of reduction embodied in Kim's more recent model, reduction is impossible unless mental properties, including qualia, can be functionalized—a question that I (and many others, including Kim) have seen reason to answer in the negative.

The Problem of Mental Causation

Mental causation tends to be a problem for physicalism. The strong emphasis on both the necessity and the sufficiency of physicalistic explanation makes it difficult to preserve conceptual space for mental phenomena to be causally effective. Epiphenomenalists, of course, accept this and consign the mental to causal irrelevance. But the majority of contemporary materialists find such a conclusion unacceptable. A dramatic statement on the importance of mental causation comes from Jerry Fodor: "If it isn't literally true that my wanting is causally responsible for my reaching, and my itching is causally responsible for my scratching, and my believing is causally responsible for my saying, . . . if none of that is literally true, then practically everything I believe about anything is false and it's the end of the world."[49] Kim responds, "One could hardly declare one's yearnings for mental causation with more feeling than this!"[50] Kim seems

---

for all mental properties—then mind-body reduction is in principle possible, if not practically feasible. This is contrary to one piece of current philosophical wisdom, the claim that functionalism, as distinguished from classic type physicalism, is a form— in fact, the principal contemporary form—of mind-body antireductionism. What I am urging here is the exact opposite—that the functionalist conception of mental properties is *required* for mind-body reduction. In fact, it is necessary and sufficient for reducibility" (*Mind in a Physical World*, chap. 4).

48. Ibid.

49. Fodor, "Making Mind Matter More," in John Heil and Alfred Mele, eds., *Mental Causation* (Oxford: Clarendon, 1993).

50. Kim, "The Nonreductivist's Troubles with Mental Causation," in *Supervenience and Mind*, p. 349.

here to view the issue as an emotive one; I should say, on the contrary, that what Fodor says is strictly and literally true—assuming that the "world" that comes to an end is the world *as humanly understood and interpreted*, not the physical universe.[51] But Kim himself is by no means indifferent to mental causation; he allows that "when I walk to the water fountain for a drink of water, my legs move in the way they do in part because of my desire for water and my belief that there is water to be had at the water fountain."[52]

The desire to preserve mental causation, and to avoid epiphenomenalism, has clearly been one of the motives behind the adoption of the identity theory (e.g., for Davidson). If mental events just *are* brain-events, and brain-events are causally effective, then mental events are causally effective.[53] But the difficulties that confront the identity theory also reopen the question of mental causation, and epiphenomenalism again becomes a live option.

As one would expect, Kim has addressed himself to this topic. The problem is set, in the first place, by an assumption Kim thinks any physicalist will embrace, namely "the causal closure of the physical domain." Causal closure is defined as follows: "*Any physical event that has a cause at time t has a physical cause at t.*"[54] This does not explicitly rule out mental causation, but it certainly makes it problematic: given the (sufficient) physical cause, what is there left for a mental cause to do?

The problem is further intensified by Kim's principle of "causal-explanatory exclusion," which states that "there can be *at most* one complete and independent causal explanation, or one fully sufficient cause, for any single event."[55] Once again, this seems to threaten any recognition of mental-to-physical causation: given that the physical explanation of any physical event is "fully sufficient," what room is there for mind to play a role in its explanation? But the exclusion principle also opens a small window of opportunity: mental causation may perhaps still play a

51. For argument on this point, see the next chapter.
52. Kim, "Myth of Nonreductive Materialism," p. 280.
53. But are they effective *in virtue of their mental characteristics?* This issue will be explored in the next chapter.
54. Kim, "Myth of Nonreductive Materialism," p. 280.
55. Kim, "Dretske on How Reasons Explain Behavior," in *Supervenience and Mind*, p. 291. (See also chap. 13, "Mechanism, Purpose, and Explanatory Exclusion," pp. 250ff.) Some special treatment will be required for cases of causal overdetermination, but overdetermination is clearly unsatisfactory as a general account of mental causation.

role, if the operation of the mental cause is not "independent," in the relevant sense, from that of the physical cause. Kim poses the problem nicely: "The delicate task is to find an account that will give the mental a substantial enough causal role to let us avoid 'the great paradox of epiphenomenalism' without infringing upon the closedness of physical causal systems."[56]

Kim has sought to exploit this window with his model of *supervenient causation*. This model is set forth as follows: "When a mental event M causes a physical event P, this is so because M is supervenient upon a physical event, P*, and P* causes P. . . . Similarly, when mental event M causes another mental event M*, this is so because M supervenes on a physical state P, and similarly M* on P*, and P causes P*."[57] More generally, "a mental event is a cause, or an effect, of another event in virtue of the fact that it is supervenient on some physical event standing in an appropriate causal relation to this event."[58]

It is clear how this proposal evades the exclusion principle: the supervenient event (and therefore the explanation involving this event as a cause) is *not independent* of the relevant physical event and its corresponding explanation, so it is not ruled out as a cause by the exclusion principle. Kim argues, furthermore, that the relationship between mental events and physical events is analogous to that between macrophysical events and their corresponding microphysical events; thus, even if someone chooses to say that mental events are "epiphenomenal," the charge has no sting: "*Epiphenomenal causal relations involving psychological events . . . are no less real or substantial than those involving macrophysical events. They are both supervenient causal relations.*"[59]

Kim's explanation of supervenient causation is admirably lucid, and for this we are in his debt. One may still question, though, whether the causal role it assigns to the mental is sufficiently robust to avoid the worries with which this inquiry began.[60] In his "Postscripts on Mental

---

56. Kim, "Epiphenomenal and Supervenient Causation," in *Supervenience and Mind*, p. 106.

57. Ibid.

58. Kim, "Myth of Nonreductive Materialism," p. 283.

59. Kim, "Epiphenomenal and Supervenient Causation," p. 107; emphasis in original.

60. Consider, for instance, the following observation by Marcelo Sabates: "We have to accept, therefore, that the supervenient causation proposal collapses into a resignation strategy: the rejection of mental causation. In this sense, the supervenient causation strategy is just a sophisticated version of epiphenomenalism" (Marcelo

Causation," Kim points out that the classical epiphenomenalists would likely accept all of the substantive assumptions involved in the supervenient causation model. (Epiphenomenalists have generally said that mental events are *caused by* physical events, not that they supervene on them, thus allowing for a time-lag between physical and mental events that supervenience denies. But Kim thinks the epiphenomenalists would have been happy to accept the supervenience relation, had this been available to them.) In view of this, Kim asks, "If 'supervenient causation' is something that even the epiphenomenalist can live with, might it not be 'causation' in name only? Is it a robust enough relation to vindicate the causal efficacy of the mental?"[61] Kim elaborates this in setting out his reasons why supervenient causation "may not be a fully adequate solution":

> *Ex hypothesi* M_1 is supervenient on, but distinct from, P_1, and P_1 is a sufficient cause of P_2 (for brevity I delete references to *instances* of these properties). But if P_1 is a sufficient cause of P_2, what causal work is there for M_1 to contribute in the causation of P_2? Granted that M_1 is supervenient on, and dependent on, P_1, and hence not an independent cause of P_1: so long as M_1 remains a distinct property not identified with P_1, we must, it would seem, still contend with the two purported causes of a single event. Given the assumption implicit in this model that fundamental causal processes occur at the physical level, the causal role imputed to M_1 in relation to an event at the physical level should strike us as something mysterious, and we should wonder what purpose could be served by this shadowy "supervenient cause" that accompanies the physical cause.[62]

Sabates, *Mental Causation: Property Parallelism as an Answer to the Problem of Exclusion* [Ph.D. diss., Brown University, 1996]).

61. Kim, "Postscripts on Mental Causation," in *Supervenience and Mind*, p. 359. This worry has not gone away for Kim. In *Mind in a Physical World*, chap. 4, sec. 6, he writes as follows: "The real bad news is that some mental properties, notably phenomenal properties of conscious experiences, seem to resist functionalization, and this means that there is no way to account for their causal efficacy within a physicalist scheme. . . . If you stay with physicalism, you come to [a] choice point: Either you retain supervenient and yet irreducible (that is, nonfunctionalizable) mental properties, say qualia, but accept their causal impotence, or you embrace mental eliminativism and deny the reality of these irreducible properties."

62. Ibid., p. 361.

Clearly, the problem of mental causation remains a vexed issue for Kim. We close this section with a remark written a few years earlier than those just cited, yet indicative of his continuing engagement with the issue: "I am not entirely certain that this supervenience solution will suffice; that is, I am not certain that anything short of the identity solution will resolve the exclusion problem."[63]

Token Identity Reconsidered

If identity is what Kim wants, perhaps he can have it at a modest cost. As noted above, one of the principal obstacles to a theory of mind-body token identity lies in Kim's own theory of events, which seems to preclude identifying mental and physical events. Nevertheless, I will suggest in this section that a small modification of this theory of events, a modification Kim himself might well accept,[64] serves to reinstate token identity as a viable option. This is so, at least, if we assume that the mind-body connection is (at least) strong supervenience.

Begin with the "species-specific bridge laws," of the form "$S_i \to (M \leftrightarrow P_i)$," that according to Kim are presupposed by Putnam's argument for multiple realizability. As we've noted, such laws are also clearly implied by the strong supervenience of the mental on the physical—a view that physicalists are hard pressed to repudiate. Now, we subject the bridge law formula to some transformations. First of all, we put it into subject-predicate notation:

$$S_I x \to (M_I x \leftrightarrow P_I x).$$

This says: If x exemplifies structure-type S_I, then it exemplifies mental state M_I if and only if it exemplifies physical state P_I. Next, we apply Importation, with the result:

$$(S_I x \;\&\; M_I x) \leftrightarrow (S_I x \;\&\; P_I x).$$

63. Kim, "Myth of Nonreductive Materialism," p. 283.

64. Kim certainly doesn't regard his theory of events as set in stone. In the preface to *Supervenience and Mind*, he writes, "In Essays 1 and 3, I formulated and argued for what is now standardly called the 'property exemplification' account of events, and I still think that it is a viable approach. However, I am now inclined to think that ontological schemes are by and large optional, and that the main considerations that should govern the choice of an ontology are those of utility, simplicity, elegance, and the like" (p. ix).

Then we separate the two conditionals, as follows:

$$[(S_Ix \mathbin{\&} M_Ix) \to (S_Ix \mathbin{\&} P_Ix)] \mathbin{\&} [(S_Ix \mathbin{\&} P_Ix) \to (S_Ix \mathbin{\&} M_Ix)].$$

Now, consider the modal status of the two conditionals. Clearly, "$(S_Ix \mathbin{\&} P_Ix) \to (S_Ix \mathbin{\&} M_Ix)$" must be considered metaphysically necessary; this much is implied by the final clause of the definition of strong supervenience. But what about the converse, "$(S_Ix \mathbin{\&} M_Ix) \to (S_Ix \mathbin{\&} P_Ix)$"? Here it seems reasonable (as was suggested above with regard to properties) to regard the basic causal laws as incorporated into the structure-type S_I. A similar structure obeying different causal laws would have different causal powers than the structure designated by S_I, and that seems sufficient reason to consider it as being a different structure. But if so, the conditional "$(S_Ix \mathbin{\&} M_Ix) \to (S_Ix \mathbin{\&} P_Ix)$" is also metaphysically necessary; an organism that (given the laws of nature that actually obtain) could exemplify the mental event M_I in some way other than by exemplifying P_I would not be an instance of the type S_I. Putting these two metaphysically necessary conditionals together, we have

$$\Box\{(S_Ix \mathbin{\&} M_Ix) \leftrightarrow (S_Ix \mathbin{\&} P_Ix)\},$$

which states that the two events, $(S_Ix \mathbin{\&} M_Ix)$ and $(S_Ix \mathbin{\&} P_Ix)$, are *metaphysically equivalent*: in *no logically possible world* does one exist without the other.

And now don't we have good reason to assert that $(S_Ix \mathbin{\&} M_Ix)$ and $(S_Ix \mathbin{\&} P_Ix)$ describe the *same event*, rather than two distinct events? The two are metaphysically indistinguishable, inseparable in all possible worlds. If we are prepared (as an earlier discussion suggested we should be) to accept metaphysical equivalence as a criterion of property-identity, then by parity of reasoning we ought to accept the metaphysical equivalence of events involving the same substance as a criterion of event-identity.[65] And since strong supervenience, with the aid of a few plausible auxiliary hypotheses,

65. In discussion, Kim suggested *there being husbands* and *there being wives* as metaphysically equivalent states of affairs that nevertheless seem to be distinct. As he noted, these are general states of affairs rather than "events" in the sense captured by his definitions. Even so, I think it may be plausible to accept the example and maintain that these are indeed merely two different descriptions of the same state of affairs, namely *there being married couples*. (If widowers were counted as husbands, and widows as wives, the statements would not be equivalent.)

supports the metaphysical equivalence of mental and physical events, it also supports the contention that mental event-tokens are identical with physical event-tokens.

At this point it may be helpful to state formally (following Kim's lead) the theory of events developed here:

> *Existence condition*: Event $[x, P, t]$ exists just in case substance x has property P at time t.

> *Identity conditions*:
> (a) If $x = y$, $P = Q$, and $t = t'$, then $[x, P, t] = [y, Q, t']$.

Following up on Swinburne's suggestions, we have also

> (b) If $[x, P, t]$ and $[y, Q, t']$ satisfy the same intrinsic description, then $[x, P, t] = [y, Q, t']$.
> (c) If P is a determinable of which Q is a determinate, then $[x, P, t] = [x, Q, t]$.

And finally,

> (d) If $\Box\{[x, P, t]$ if and only if $[x, Q, t]\}$, then $[x, P, t] = [x, Q, t]$.

It may be useful at this point to compare the theory of events set out above with Davidson's. A first point to notice is that the criterion of event-identity is still narrower than Davidson's. It is quite plausible to argue that if $[x, P, t]$ and $[y, Q, t']$ are metaphysically indistinguishable, they have all their causes and all their effects in common. But the converse does not seem to hold, or at least it is far from evident that it holds. So it seems that our present theory will allow for distinct events that on Davidson's theory will be identical.

The more important difference, though, concerns the way in which the present theory avoids the incoherence that plagues Davidson's theory. That incoherence, it will be remembered, takes the form of a circularity: events are identified by the causal relations into which they enter, and causal relations are identified as relations between events. The present theory avoids this by initially identifying events as is done by Kim, in terms of the substance, property, and time involved in each event. It then proceeds to reduce the excessive multiplicity of events as given by Kim's theory, by stating various conditions under which differently described events are nonetheless identical.

It may seem, however, that this procedure opens the theory up to a charge of inconsistency. If we begin with Kim's way of identifying events, don't we have to stick to that same approach in deciding which events are distinct? So isn't the present proposal just a cobbled-together monstrosity, blending together incompatible ideas in order to arrive at a desired conclusion?

I think such a reaction would be a mistake. Certainly, Kim's theory initially identifies events by describing them in a certain way. That, after all, is what *any* theory of events is bound to do. But it is surely part of our ordinary conception of events (as of other sorts of entities) that the same event can be described in different ways, and so different (and logically nonequivalent) descriptions may pick out the same individual event. Kim's own theory allows for this, in saying that "$[x, P, t] = [y, Q, t']$ just in case $x = y$, $P = Q$, and $t = t'$." What we have done, is merely to extend this to a few additional sorts of cases. I believe the two modifications suggested by Swinburne possess great intuitive plausibility. Surely Brutus's stabbing of Caesar is identical with the event that caused Caesar's death, even though the latter (non-intrinsic) description mentions neither the subject of the action nor the action-type involved. Surely, also, it is the case that Swinburne's walking slowly at noon is the same event as Swinburne's walking at noon, provided Swinburne does walk slowly.[66] Similarly, I hope and believe that upon reflection most of us will find it extremely plausible that metaphysically equivalent events are in fact identical. The general point, however, should be abundantly clear: there is absolutely no reason to suppose that the descriptions by which we initially identify events must correspond one-to-one with distinct events, so that no two descriptions can pick out the same event.

To sum up, I believe the approaches to the mind-body relationship sketched out in the latter part of this chapter are good candidates for coherent, reasonably plausible physicalist theories. Strong supervenience, as we have seen, can reasonably be regarded as the minimum claim any theory of the mind-body relation must make in order to qualify as physicalist or materialist.[67] The argument for token-identity no doubt involves a few more debatable assumptions, but it is my hope that a good many philosophers inclined to physicalism will find those assumptions conge-

66. For evidence that Kim might accept this, see "Postscripts on Mental Causation," pp. 361–62.

67. We are of course thinking here of *non-eliminativist* theories.

nial.[68] Both, then, can stand as reasonable representatives of a type of mind-body theory that is neither too weak to qualify as a respectable physicalism nor so strong as to be either incoherent or immediately implausible. Whether such theories will in the end deserve our acceptance is the topic of succeeding chapters.

68. Kim's own present view is a development of the "multiple-type physicalism" described in "Postscripts on Mental Causation," pp. 362–67. I would find the token-identity theory proposed here preferable to multiple-type physicalism, on the ground that it retains mental properties such as qualia as intrinsic, first-order properties—a view for which, as we've seen, Kim has expressed some sympathy. Kim objects to the token-identity theory in that it "is entirely silent on the important insight concerning psychophysical type-type relations that is contained in the multiple realization thesis" (ibid., p. 365). Kim appears to have Davidson's theory in mind here; the present proposal, in view of the "species-specific bridge laws" and their derivatives, certainly does not ignore psychophysical type-type relations.

CHAPTER THREE

Why the Physical Isn't Closed

Descartes thought that whereas reflex actions and vital processes could be accounted for physiologically, the higher rational processes of human beings, and in particular the use of language, required a nonphysical explanation. It is now generally taken for granted that Descartes was wrong about this, with the existence of computers providing a ready counterexample. I believe, on the contrary, that Descartes was substantially correct, and this chapter will develop arguments in support of this contention. If successful, these arguments will constitute a general refutation of materialism or physicalism as a theory of the mind.

Claims as dramatic as this require rigorous argument, and such argument must have a well-defined target. Because of this, the first section of the chapter will bring into sharp focus the precise aspect of materialism that, it will be argued, makes a materialistic explanation of rational thought impossible. It will turn out that the "causal closure of the physical domain"—given a specific understanding of "physical"— represents precisely the feature of physicalism which renders physicalism incapable of explaining rational thought processes; thus the title of this chapter. Then we will review some arguments in support of this claim.

SUPERVENIENCE, CAUSAL CLOSURE, AND MECHANISM

In the previous chapter we developed a specific version of physicalism which, it is hoped, is both reasonably plausible and strong enough in its claims to qualify as a genuine materialism. In the process two different claims emerged, each of which may well be regarded as characteristic of materialism as a whole. One such claim is that the mental domain is supervenient on the physical domain. Jaegwon Kim's assertion about the importance of supervenience is unusually strong: "Acceptance or rejection of the supervenience of the mental on the physical leads to the most basic division between theories of the mind-body relation: theories that accept psychophysical supervenience are fundamentally materialist, and those that reject it are fundamentally antimaterialist. This difference seems philosophically more basic and more significant than the usual classification of mind-body theories as 'monist' or 'dualist.' "[1] I believe Kim is right about this. What supervenience guarantees is that *the mind cannot vary independently of the body*, and this seems to be an inescapable implication of the physicalist claim that the physical facts determine all the facts. Absent this, we could have a mental domain containing facts that are as they are independently of what happens to a person's body, and no materialist worth her salt will countenance that.

Another doctrine that has a claim to be characteristic of materialism or physicalism as a whole is the *causal closure of the physical domain*. Kim, indeed, terms this "a principle that seems minimally required of any serious form of physicalism."[2] But what exactly is this closure principle? Kim explains it as follows: "*any physical event that has a cause at time t has a physical cause at t*."[3] As one would expect from Kim, the definition is carefully

1. Jaegwon Kim, "Concepts of Supervenience," in *Supervenience and Mind: Selected Philosophical Essays* (Cambridge: Cambridge University Press, 1991), pp. 56–57. Kim points out that "mind-body supervenience is consistent with a host of classic positions on the mind-body problem," including emergentism, functionalism, type physicalism, epiphenomenalism, and even some forms of dualism. He suggests that "it is useful to think of the diverse mind-body theories as offering competing explanations of mind-body supervenience" ("The Mind-Body Problem after Fifty Years" [lecture delivered at the Royal Institute of Philosophy, London, October 28, 1996], sec. 2).

2. Kim, "Dretske on How Reasons Explain Behavior," in *Supervenience and Mind*, p. 290.

3. Kim, "The Myth of Nonreductive Materialism," in *Supervenience and Mind*, p. 280.

crafted. It doesn't claim that *all* physical events have physical causes; given quantum mechanics, it would be rash to assert dogmatically that there can't be uncaused events. Nor does it say that *only* physical events can be causes. On the contrary, Kim wants to leave conceptual space for mental causation, though (as we saw in the last chapter) he has considerable difficulty in doing so. Kim goes on to explain:

> This is the assumption that if we trace the causal ancestry of a physical event, we need never go outside the physical domain. To deny this assumption is to accept the Cartesian idea that some physical events need nonphysical causes, and if this is true there can in principle be no complete and self-sufficient physical theory of the physical domain. If the causal closure failed, our physics would need to refer in an essential way to nonphysical causal agents, perhaps Cartesian souls and their psychic properties, if it is to give a complete account of the physical world.[4]

As he elsewhere says, "If this is what you are willing to embrace, why call yourself a 'physicalist'?"[5] Once again, it seems to me that Kim is justified in his insistence on the importance of causal closure for any legitimate physicalism or materialism.[6]

So we have two different principles, each of which has a strong claim to being a bedrock requirement for materialism. One is led to ask about the relationship between the two. The correct answer seems to be the following: The causal closure of the physical domain is a requirement for all materialists without exception; as Kim rightly says, no materialist can countenance the idea that some physical events require nonphysical causes. This requirement holds also for eliminative materialists, who don't admit that there are any mental events. But if one is in some minimal sense a realist about the mental, so that mental events do exist, the supervenience principle comes into play to insure that these mental facts are determined

4. Ibid.
5. Kim, "The Nonreductivist's Troubles with Mental Causation," in *Supervenience and Mind*, p. 356.
6. Interestingly, the significance of causal closure is confirmed independently by the dualist Karl Popper, who writes, "I call this the physicalist principle of the closedness of the physical World 1. It is of decisive importance, and I take it as the characteristic principle of physicalism or materialism" (Karl R. Popper and John C. Eccles, *The Self and Its Brain: An Argument for Interactionism* [New York: Springer International, 1977], p. 51).

by physical facts. In short: causal closure is a requirement for all materialists without exception, while supervenience must be affirmed by materialists who are realists about the mental.

We still have not reached the end of our quest. The closure principle
says that all physical events have physical causes, yet I have undertaken
to show that there are certain events—specifically, human rational
thought processes—that can't possibly be explained in terms of physical
causes. What, then, are the limitations of physical causes that make such
explanations impossible? And more generally, what counts as "physical"
here?

The task of framing a general definition of the physical realm is not a
trivial one. According to Descartes, spatial extension is the defining characteristic of the physical—if something is physical it is extended, and vice
versa. After reviewing several other proposed definitions, Charles Taliaferro pronounces this "among the most promising."[7] But unless one presupposes Descartes's entire metaphysical scheme, this definition is problematic. Some philosophers have posited minds, or mindlike entities,
essentially similar to Cartesian souls, except that they are spatially located.
(Stewart Goetz and Philip Quinn are two contemporaries who have made
such proposals.)[8] It is unlikely in the extreme, however, that any contemporary materialist is going to acknowledge these as physical entities just
because they are assigned a position in space. Taliaferro, recognizing that
some clearly nonphysical items might be spatial, observes that this would
leave spatiality as a necessary condition, even if not a sufficient condition,
for physicality.[9] This, however, hardly suffices to delineate the scope of the
physical realm. With regard to a general conception of the physical, Taliaferro states, "By 'physical' I will mean those things that are described and
posited in mainstream current physics, or things like them, things that
can be publicly observed, or things composing publicly observable objects, and things that are spatially located."[10] Such a disjunctive criterion
may be usable for certain purposes, but it is too loose (and potentially

7. Charles Taliaferro, *Consciousness and the Mind of God* (Cambridge: Cambridge
University Press, 1994), p. 98.
8. See Philip Quinn, "Tiny Selves: Chisholm on the Simplicity of the Soul," in
Lewis Edwin Hahn, ed. *The Philosophy of Roderick M. Chisholm* (Peru, Ill.: Open
Court, 1997), pp. 55–67; Stewart Goetz, "Souls and Spatial Location," unpublished.
9. Taliaferro, *Consciousness and the Mind of God*, p. 101. The specific objects Taliaferro has in mind are visual images.
10. Ibid., p. 104.

open-ended) to serve in the present context, where we are seeking to determine the minimum requirements for physicalistic explanation.[11]

Aware of the difficulties involved in framing an abstract definition, Kim suggests that we should "explain 'physical' by reference to current theoretical physics."[12] But this can't be quite right either. The changes in physics over the past several centuries alert us to the likelihood that new objects may be introduced in the future that are significantly unlike those contained in current physics; nevertheless, we would surely want to count these new objects as "physical." But if we stretch the meaning of "physical" to include the denizens of some future physics, or of some ideal physics whether or not it ever actually exists, then it's difficult to know what if anything is being ruled out. Mindful of such difficulties, Alan Donagan concluded that "understood as the doctrine that nothing happens that is not causally explicable in terms of the natural sciences as they now are, materialism is certainly false; and . . . understood as the doctrine that nothing happens that is not causally explicable in terms of ideal natural science, it is something we know not what."[13]

All this raises the question, is there some way to understand "physical" that will be sufficiently flexible to allow for future scientific progress, yet will constrain the meaning enough that it doesn't become too nebulous for useful discussion? I believe there is indeed a way to do this. As an initial statement, I propose the following: *All physical causation and physical explanation must be mechanistic.* To be sure, this invites questions about the meaning of "mechanistic" similar to those already encountered for "physical." "Mechanistic" suggests "mechanical," but no one thinks the universe is literally a machine. By seventeenth-century standards, which tended to require contact action, Newtonian gravitation was nonmechanistic. By Newtonian standards, Maxwell's theory of electromagnetism was nonmechanistic. And quantum *mechanics* is probably the least "mechanical" physical theory to date. So we ask again: What does this proposal rule out?

I believe it is possible to assign a meaning to "mechanistic" which is

11. In fairness, it should be pointed out that Taliaferro's primary objective here is not to craft an airtight definition of the physical, but rather to show that the more plausible definitions all presuppose a prior understanding of the mental.

12. Kim, "Nonreductivist's Troubles with Mental Causation," p. 340.

13. Alan Donagan, "Can Anybody in a Post-Christian Culture Rationally Believe the Nicene Creed?" in Thomas P. Flint, ed., *Christian Philosophy* (Notre Dame, Ind.: University of Notre Dame Press, 1990), p. 107.

broad enough to accommodate all present (and likely future) physical science, yet narrow enough to impose serious constraints on what can count as "physical." Furthermore, this meaning corresponds fairly closely to the way "mechanistic" has been used in some recent discussions. Here is the proposal: *Mechanistic causation and mechanistic explanation are fundamentally nonteleological.* That is to say: in any instance of mechanistic causation, the *proximate cause* of the effect does not involve a goal, objective, or *telos*; rather, it consists of some disposition of masses, forces, and the like. Similarly, a mechanistic explanation does not say why an event occurred in terms of some goal that was being reached or some purpose or function that was being served; rather, it appeals to antecedent conditions involving only nonpurposive, nonintentional entities. To be sure, an event caused mechanistically can also have a true teleological explanation: a thermostat turns the furnace on and off *in order to* maintain a constant temperature, and the fact that it works that way has a cause in the human *desire* for a comfortable environment. But the *proximate cause* of the thermostat's function has no reference to such purposes and desires: the thermostat turns on the furnace because a certain strip of metal, cooled by the ambient air, became bent in such a way as to close an electrical circuit.

Socrates captured the point exactly in his complaint about the physics of Anaxagoras:

One day, however, I heard someone reading from a book he said was by Anaxagoras, according to which it is, in fact, intelligence that orders and is the reason for everything. Now this was a reason that pleased me; it seemed to me, somehow, to be a good thing that intelligence should be the reason for everything. . . . Reckoning thus, I was pleased to think I'd found, in Anaxagoras, an instructor in the reason for things to suit my own intelligence. . . .

Well, my friend, those marvellous hopes of mine were dashed; because, as I went on with my reading, I beheld a man making no use of his intelligence at all, nor finding in it any reasons for the ordering of things, but imputing them to such things as air and ether and water and many other absurdities.[14]

Socrates, not unreasonably, abandoned physics and pursued instead inquiries better suited to answer his questions about what is best and how it

14. Plato, *Phaedo* 97–98, trans. and ed. David Gallop (Oxford: Oxford University Press, 1993), pp. 53–54.

may be pursued. But ancient physics, with the exception of the atomists, did on the whole tend to give some encouragement to those who sought from it an account of "the best" and of the reasons for things—most enduringly, in Aristotle's doctrine of final causes. And it was the expulsion of final causes from physics by Descartes and Galileo that marked what was perhaps the most decisive break between ancient and modern natural science.[15] A future revision of physics that would reverse this shift would be astonishing.

Having said this, our present concern is not to legislate the future of science but rather to stipulate our present use of the terms "physical" and "mechanistic." If some future revision of physics should, contrary to all expectation, reintroduce teleology into (what is then called) physical explanation, all the issues discussed in this chapter would have to be reopened. In the meantime, what we have to consider is this: Can explanations that are mechanistic, in the sense described, do justice to human action and in particular to human rationality?

THE ARGUMENT FROM REASON

In the late 1960s there took place in the pages of *The Philosophical Review* a discussion between Norman Malcolm and Alvin Goldman. In his paper, "The Conceivability of Mechanism," Malcolm argued that there is an incompatibility between "purposive explanations" of human actions and complete explanations of those same actions in nonintentional, neurophysiological terms. Once a complete mechanistic explanation has been given, no room is left for the reasons one might have for an action to play a role in bringing it about that one performs the action. If this is so, then the doctrine that all actions have complete neurophysiological explanations entails that no one *ever* does anything because she has reason to do so—that is, no one ever performs an intentional action. Mechanism is inconceivable, according to Malcolm, not because it is self-contradictory (it is not), but rather because "the absurdity lies in the human act of asserting the doctrine. The occurrence

15. According to David L. Hull, "Historically, explanations were designated as mechanistic to indicate that they included no reference to final causes or vital forces" (*The Cambridge Dictionary of Philosophy*, s.v. "mechanistic explanation").

of this act of assertion is inconsistent with the content of the assertion."[16]

Goldman's reply took the form of arguing that, under certain special conditions, there can indeed be a true purposive explanation and a true (and complete) mechanistic explanation for the same action. The special conditions involved are explained in terms of the notion of "simultaneous nomic equivalents," which he defines as follows:

> Suppose there is a (contingent) law saying that for any object of kind H and any time t, the object has property ϕ at t if and only if it has property ψ at t. Then if a particular object a has properties ϕ and ψ at a particular time t_1, I shall say that a's having ϕ at t_1 is a "*simultaneous nomic equivalent*" of a's having ψ at t_1.[17]

The application is as follows: Suppose one's mental state and one's neurophysiological state at a particular time are simultaneous nomic equivalents. Then if one of these states can serve as an explanation for a given event—that is, as an antecedent condition from which the event's occurrence can be inferred—so can the other. So, given this special assumption (one materialists are very likely to make), reasons-explanations and mechanistic explanations are compatible after all.

Kim, in commenting on this discussion, awards partial credit to both Malcolm and Goldman. In support of Malcolm, Kim endorses a "principle of explanatory exclusion," which states that "no event can be given more than one *complete* and *independent* explanation."[18] And if we accept certain additional assumptions Malcolm makes, it follows that the two types of explanations are indeed incompatible. On the other hand, if two events are simultaneous nomic equivalents in Goldman's sense, the explanations are *not* independent and so may be compatible after all. This, however, does not conclusively settle the issue in Goldman's favor. That two events are simultaneous nomic equivalents does guarantee that whatever is explained by one of them will be explained by the other, *if* we ac-

16. Norman Malcolm, "The Conceivability of Mechanism," *Philosophical Review* 77 (1968): 67–68.

17. Alvin I. Goldman, "The Compatibility of Mechanism and Purpose," *Philosophical Review* 78 (1969): 473.

18. Kim, "Mechanism, Purpose, and Explanatory Exclusion," in *Supervenience and Mind*, p. 239.

cept a Hempelian account of explanation according to which the explanans is any previous event(s) that nomologically implies the explanandum. But if we are looking for a *causal* explanation, it is by no means guaranteed that both of the simultaneous nomic equivalents *cause* the event to be explained. If we replace "simultaneous nomic equivalents" with the notion of supervenience, we see that this becomes the problem of supervenient causation, over which Kim has had, and continues to have, genuine worries. These questions form the springboard from which Kim launches into a general discussion of the "possibility of *multiple explanations of a single explanandum*, and the relationship between two distinct explanatory theories covering overlapping domains of phenomena."[19]

My interest in this discussion is not so much in these further considerations as in an assumption that Malcolm, Goldman, and Kim all seem to have in common. This is the assumption that *if mind-body identity is accepted, the problem of multiple explanations is solved.*[20] As Kim says, "On the identity view, there is here only one cause of [an event], not two whose mutual relationship we need to give an account of. As for explanation, at least in an objective sense, there is one explanation here, and not two."[21] We noted in the last chapter that Kim is himself attracted to "the identity solution," though he reluctantly concludes that this solution is not available to him.

It was argued in the previous chapter that Kim can after all have mind-body identity, at the cost of a modest (and fairly plausible) revision of his criterion of event-identity. But now I want to say that *mind-body identity does not resolve the physicalist's problem with multiple, incompatible explanations.* The reason for this is illustrated by a passage from Kim's critique of Davidson:

Davidson's anomalous monism fails to do full justice to psychophysical causation in which the mental *qua mental* has any real causal role to play. Consider Davidson's account: whether or not a given event has a mental description (optional reading: whether it has a mental characteristic) seems entirely irrelevant to what causal relations it enters into. Its causal powers are wholly determined by the physical description or characteristic that

19. Ibid., pp. 237–38.

20. See Malcolm, "Conceivability of Mechanism," pp. 53–54; Goldman, "The Compatibility of Mechanism and Purpose," p. 478.

21. Kim, "Mechanism, Purpose, and Explanatory Exclusion," p. 248.

holds for it; for it is under its physical description that it may be subsumed under a causal law.[22]

It seems to me that this is entirely correct—though the result is ironic for Davidson, who argued for mind-body identity precisely on the grounds that this was the only way to secure a causal role for mental states. And they do have such a role, but only in a way that abstracts completely from their specific character *as mental states*, and as mental states *of such-and-such a kind*.

What is surprising, though, is that Kim doesn't seem to see that the criticism is equally damaging to his own view. It's true that, on either a supervenience view or the kind of token-identity view that Kim might be able to accept, there are psychophysical lawlike connections of a sort that Davidson apparently denies.[23] But this in no way obscures the fact that the mental events have the causal powers they do *only in virtue of their physical characteristics*; it remains true that "whether or not a given event has a mental description . . . seems entirely irrelevant to what causal relations it enters into. Its causal powers are wholly determined by the physical description or characteristic that holds for it; for it is under its physical description that it may be subsumed under a causal law." For consider: each mental event is either identical with or supervenient on a physical event. By hypothesis, the physical event in question has a complete causal explanation in terms of previous events *with which it is connected according to the laws of physics*. (That is implied by the causal closure of the physical domain.) Similarly, each such event has whatever causal powers it has

22. Kim, "Epiphenomenal and Supervenient Causation," in *Supervenience and Mind*, p. 106.

23. I say "apparently," because it is hard to be sure exactly what is excluded by Davidson's denial of psychophysical laws. There is a passage in "Mental Events" in which he expresses a willingness to accept a version of mind-body supervenience:

> Although the position I describe denies there are psychophysical laws, it is consistent with the view that mental characteristics are in some sense dependent, or supervenient, on physical characteristics. Such supervenience might be taken to mean that there cannot be two events alike in all physical respects but differing in some mental respect, or that an object cannot alter in some mental respect without altering in some physical respect ("Mental Events," in Donald Davidson, *Essays on Actions and Events* [Oxford: Clarendon, 1980], p. 214).

Sorting out the precise interpretation of Davidson's position is beyond the scope of the present discussion.

solely in virtue of its physical characteristics, such powers being exercised, once again, according to the physical laws. No causal role for the mental characteristics as such can be found. And if this criticism is telling against Davidson, which it is, it is also heavily damaging to Kim's own theory.

At this point, I wish to transpose the discussion from the context of *reasons for action*, and their causal role in bringing about behavior, to the context of *reasons for belief*, and the role they play in the process of *rational inference* by which new beliefs are accepted. It's evident that the two contexts are closely related, and in fact all the points that have been made in the discussion about reasons for action apply with equal force to reasons for belief. Thus, one can argue against both Kim and Davidson— and in fact, against *any* physicalist view that maintains the causal closure of the physical domain—that "whether or not a given event has a mental description" (for instance, as the acceptance of a proposition which constitutes a good reason for some other proposition) "seems entirely irrelevant to what causal relations it enters into" (for instance, to what other beliefs a person comes to accept as a result). To put it more plainly, *On the assumption of the causal closure of the physical, no one ever accepts a belief because it is supported by good reasons.* To say that this constitutes a serious problem for physicalism seems an understatement.

Before pursuing this further, it will be well to say a word about the nature of the argument that is being presented. The previous paragraph may suggest what has been termed a Skeptical Threat argument, one that aims "to raise skeptical doubts about the validity of reasoning, and then to argue that such doubts can be resolved only if [physicalism] is denied."[24] Skeptical Threat arguments have about them an air of paradox (like that of skeptical arguments generally) that tends to keep them from being taken seriously. It may be replied, in the spirit of G. E. Moore, that we have overwhelmingly strong reasons for acknowledging the "validity of reasoning"—that is, for acknowledging that people do sometimes reach conclusions because of good reasons they accept, and that they are rational in doing so—and that, therefore, any argument to the contrary must be based on a mistake or trick of some kind. And such a reply would, I think, have considerable force. In any case, it is not my intention here to advise physicalists that they ought to give up their belief in

24. Victor Reppert, "The Lewis-Anscombe Controversy: A Discussion of the Issues," *Christian Scholar's Review* 19 (September 1989): p. 37.

reasoning and inference as rational processes that are helpful in leading us to the truth.[25]

The Argument from Reason is best understood, however, not as a Skeptical Threat argument but rather as a Best Explanation argument. Such an argument begins by *assuming* the validity of reasoning and then asking how that validity can be accounted for, given the assumptions of physicalism. The argument then asks, "not *whether* reason is valid, but whether, in a [physicalistic] world, one can *account* for the fact that it *is* valid."[26] If it should turn out that the physicalist *cannot* account for this—if, in fact, key assumptions of physicalism make it impossible to give an acceptable account of rational inference—this failure will constitute an extremely serious objection to physicalism, if not an outright refutation.

I have claimed that, on the assumption of the causal closure of the physical, no one ever accepts a belief because it is supported by good reasons. Since this assertion is key to the argument, some further discussion is in order. What we have to consider is the relationship between the physicalistic explanation of a person's holding a belief—"She believes so-and-so because of such-and-such antecedent physical conditions"—and the rational explanation of that same belief—"She believes so-and-so because she sees that it is supported by sound reasons." By hypothesis, the physical causes are sufficient, under the given conditions, to produce the belief in question. There can be no question, on the other hand, of the reasons for the belief being *by themselves* sufficient to produce the belief. For the reasons to give rise to the belief, the person's cognitive apparatus has to be in working order, and this includes a vast number of extremely complex circumstances that are quite distinct from the possession of the reasons in question.

But is the possession of good reasons *necessary*, under the given circumstances, for the production of the belief in question? What we have to evaluate is the following pair of counterfactual conditionals:

25. The possibility exists, however, that such considerations as these may lead someone to have genuine skeptical doubts, somewhat as Darwin wondered whether our minds, descended as they are from the lower animals, could be trustworthy. One would hope the result would be a questioning of the physicalist assumption from which the doubts arise, but this can't be guaranteed.

26. Reppert, "Lewis-Anscombe Controversy," p. 37.

(a) She would have accepted the belief if she had not seen that it was supported by good reasons.

(b) She would not have accepted the belief if she had not seen that it was supported by good reasons.

I submit that, in the absence of further information, neither of these counterfactuals can be evaluated as true. Following John Pollock, we assume that a counterfactual conditional is true if and only if the consequent is true in all those worlds minimally changed from the actual world in which the antecedent is true. Would a world minimally changed from the actual world in which she doesn't see that her belief is supported by good reasons, be one in which she would not accept that belief? No doubt there are a number of different ways in which the world could be changed just enough to satisfy the antecedent of the conditional; in some of these she accepts the belief while in others she doesn't. And there is no basis for saying that those in which she doesn't accept it are less changed from the actual world than those in which she does—or *vice versa*.[27] We conclude, then, that *(a)* and *(b)* are both false; what is true is

(c) If she had not seen that the belief was supported by good reasons, she might have accepted the belief, but it's also the case that she might not have accepted it.[28]

Consider, on the other hand:

(d) She would have accepted the belief if the antecedent conditions were not sufficient (as determined by the laws of physics) for her accepting it.

(e) She would not have accepted the belief if the antecedent conditions were not sufficient (as determined by the laws of physics) for her accepting it.

27. It is generally agreed that the causal laws which obtain are important in determining similarity relationships between possible worlds. But in our physicalistic world the principles of sound reasoning have no causal relevance, thus they can't be used to support the conclusion that worlds in which the subject rejects the belief are closer than those in which she accepts it.

28. In symbols: "~R $\Box\!\!\rightarrow$ B" and "~R $\Box\!\!\rightarrow$ ~B" are both false; whereas "~R $\Diamond\!\!\rightarrow$ B" and "~R $\Diamond\!\!\rightarrow$ ~B" are both true.

Here we can state unambiguously that *(d)* is false and *(e)* is true; the fact that one, and not the other, is in accord with the physical laws means there is no question that worlds in which she does not accept the belief are closer to the actual world than the ones in which she does. All of this merely restates, in the language of counterfactual conditionals, what should by now be obvious: *In a physicalistic world, principles of sound reasoning have no relevance to determining what actually happens.*

Because the matter is so crucial, I am going to risk excess by restating the point once more, this time in terms of possible worlds. In order to identify the possible worlds we want to consider, note again the final clause in the definition of strong supervenience: "necessarily, if any y has [physical property] G, it has [mental property] F." If "necessarily" here is understood as physical necessity, identifying the relevant world is easy: consider a possible world that is physically exactly similar to the present world, but in which the natural laws establishing psychophysical connections do not obtain. *In such a world all the physical facts, and with them the entire physical course of events, are exactly as in the actual world: the complete absence of mentality makes no difference whatever.* Similarly, we may consider a possible world physically identical with the actual world, but in which mental properties are redistributed in as bizarre a fashion as one might wish: this world is still indistinguishable from our own in all physical respects. Could there be a more dramatic demonstration of the fact that, given the closure of the physical, mental facts are irrelevant to the physical course of events?

Suppose, however, "necessarily" in the definition of supervenience is understood as metaphysical necessity. This embodies the idea (for which I've expressed some sympathy) that the natural laws that obtain are expressions of the essential causal powers of the kinds of objects that exist, so that in no possible world do *those very same objects* exist governed by different natural laws. This means we can't simply cancel the psychophysical connections while leaving the rest of the actual world unchanged. Instead, we proceed as follows: take a world consisting of objects *exactly similar* to the objects of our own world, *except with regard to the psychophysical connections that obtain.* For reasons that should be evident, we will designate this as the *physically equivalent zombie-world* to our own world. In the zombie-world matter will not consist of protons, neutrons, electrons, etc., but rather of zombie-protons, zombie-neutrons, zombie-electrons . . . The zombie-electrons will not have the properties of mass, charge, and spin but rather zombie-mass, zombie-charge, and zombie-spin. Such a world will be similar to the "mindless world" described in the previous paragraph in every respect but one: it will not contain the identical objects, organisms, etc. that exist in our world (and in the mindless world) because those objects consist of ordinary matter and not of zombie-matter. But the zombie-world is *physically equivalent* to both the mindless world and the actual world; all three worlds are identical in all physically observable respects. Once again, we have a dramatic demonstration of the fact that neither in the zombie-world, nor in the mindless world, nor in the actual world *given the assumption of the causal closure of the physical*, do principles of rational inference play any role whatever in determining what happens. And in the actual world (which is not mindless), the principles of inference play no role, given causal closure and supervenience, in determining what beliefs people come to accept.

One way in which a physicalist might respond here is by questioning the assumption that good reasons and principles of rationality need to be thought of as causally relevant to what happens in the world. Wittgensteinians often adopted the stance that reasons-explanations and causal explanations belong to different language-games and so do not conflict with each other.[29] And Kim recommends as "well worth exploring" the idea that rationalizing explanation is "a fundamentally *noncausal* mode of understanding actions," so that "a rationalizing explanation is to be viewed as a *normative assessment* of an action in the context of the agent's relevant intentional states."[30]

Whatever its merits in general, Kim's suggestion is singularly unpromising in its application to the relation between reasons and beliefs. To see this, the reader is asked to reflect on the way she goes about assessing an argument of moderate complexity. I presume she begins by reflecting on the premises of the argument—are they propositions she believes, or at least considers reasonably plausible? She then considers carefully the logical connections that are alleged to obtain between the premises and the conclusion—do the premises indeed provide support for the conclusion, and if so does the support amount to deductive validity, or is there

29. Elizabeth Anscombe in effect took this line in criticizing C. S. Lewis's presentation of the argument. (See Reppert, "Lewis-Anscombe Controversy," p. 44.) But as Reppert points out, "the Wittgensteinian point of view does not lend itself very easily to the kind of ontological restrictiveness that is built into naturalism" (ibid.).

30. Kim, "Mechanism, Purpose, and Explanatory Exclusion," p. 240 n. 4.

some lesser degree of support? Are there ambiguities in the argument which might undermine the soundness of the inference? Sometimes these questions are assessed in the light of specific, explicitly formulated principles of logic and argument; at other times she relies on a more intuitive grasp of the particular argument at hand. If the assessment is favorable, she accepts the conclusion, either tentatively or with considerable firmness, depending on the particulars of the case. If she is skillful in carrying out such assessments, she is said to possess "good logical insight," an intellectual virtue which is prized, in part, for the specific reason that it enables one to reach good, well-justified conclusions about the arguments one encounters.[31] The entire process makes no sense at all, except on the assumption that a person's awareness of reasons and her knowledge and application of principles of rationality *make a difference to the conclusions that are accepted.*

Kim's suggestion, as applied to the relation between reasons and beliefs, is not only implausible; it is also futile. For surely those who would argue that principles of rationality serve the purpose of a "normative assessment" of our reasoning would allow that these principles *can in fact be used* in making such an assessment. But of course, *such a normative assessment of a piece of reasoning is itself also an example of the kind of reasoning that is being assessed.* (Note that the example of reasoning described above involved precisely the examination of an already formulated argument.) Are good reasons, and the principles of sound reasoning, allowed to be causally effective in determining the outcome of the assessment process? Or is some other account to be given of how the process goes? In any case, whatever answer is given here could equally well have been given in the first place; the move to the level of "normative assessment" changes nothing.

A final escape route for the physicalist might be the adoption of a thor-

---

31. Consider the following from Reppert: "If you were to meet a person, call him Steve, who can argue with great cogency for every position he holds, you might on that account be inclined to consider him a very rational person. But suppose it were to turn out that on all disputed questions Steve rolls dice to fix his beliefs permanently, and uses his reasoning skills only to generate the best available arguments for those beliefs selected in the above-mentioned random method. I think that such a discovery would prompt you to withdraw from him the honorific title "rational." Clearly the question of whether or not a person is rational cannot be answered in a manner that leaves entirely out of account the question of how his beliefs are produced and sustained" ("Lewis-Anscombe Controversy," p. 46).

oughly externalist view of justification. What determines the justification of a belief, on this view, is not internal cognitive processes such as were described above, but rather one simple question: was the belief produced, and is it sustained, by a reliable belief-forming process? If it was, then no further questions about justification—such as those raised by the Argument from Reason—need be asked.

It is of course true that a belief, in order to be justified, needs to have been formed and sustained by a reliable epistemic practice. But, in the case of rational inference, what is this practice supposed to be? The reader is referred, once again, to the description of a reasoning process given a few paragraphs back. Is this not, in fact, a reasonably accurate description of the way we actually view and experience the practice of rational inference and assessment? It is, furthermore, a description which enables us to understand why in many cases the practice is highly reliable—and why the reliability varies considerably depending on the specific character of the inference drawn and also on the logical capabilities of the epistemic subject. And, on the other hand, isn't it a severe distortion of our actual inferential practice to view the process of reasoning as taking place in a "black box," as the externalist view in effect invites us to do? Epistemological externalism has its greatest plausibility in cases where the warrant for our beliefs depends crucially on matters not accessible to reflection—for instance, on the proper functioning of our sensory capacities. Rational inference, in contrast, is the paradigmatic example of a situation in which the factors relevant to warrant *are* accessible to reflection; for this reason, examples based on rational insight have always formed the prime examples for internalist epistemologies.

There is also this question for the thoroughgoing externalist: How are we supposed to satisfy ourselves as to which of our inferential practices *are* reliable? By hypothesis, we are precluded from appealing to rational insight to validate our conclusions about this. One might say we have learned to distinguish good reasoning from bad by noticing that good inference-patterns generally give rise to true conclusions, whereas bad inference-patterns often give rise to falsehood. (This of course assumes that our judgments about particular facts, especially facts revealed through sense perception, are not in question here—an assumption I will grant for the present.) But this sort of "logical empiricism" is at best a very crude method for assessing the goodness of arguments. There are plenty of invalid arguments with true conclusions, and plenty of valid arguments with false conclusions. There are even good inductive arguments *with all true premises* in which the conclusions are false. These are just the sort of distinctions which the science of logic exists to help us with; basing this science on the kind of ham-fisted empiricism described above is a hopeless enterprise.

Inevitably, sooner or later in this discussion the suggestion will be made that, after all, evolution and natural selection are bound to have shaped us in such a way that our cognitive processes are reasonably effective in arriving at the truth. Of course, all of us now alive are the heirs to this selection process, so the prospects for making fine discriminations on this basis don't seem very promising! But "evolutionary epistemology" will be the theme for the final section of this chapter.[32]

MECHANISM AND DARWINIST EPISTEMOLOGY

For humans in this post-Darwin era, there is a tight link between evolution and rationality. In understanding what we take to be our endowment of reason, we can no longer appeal (so it is commonly thought) to the wisdom of a beneficent Creator. We can't attribute it to a demiurge, or to a World Spirit, or even to the programmers at Microsoft. If rationality is something we've got, evolution must have given it to us.

Philosophers have not been slow to provide the needed explanation. William James's pragmatism had at its core the situating of human cognition within the evolutionary context—thinking becomes part of the struggle to survive and flourish. Karl Popper, after disparaging the philosophical importance of evolution early in his career, consumed an unac-

---

32. The type of argument developed in this section has a substantial history, though recently it has been neglected. Descartes's remarks, quoted at the beginning of the chapter, anticipate the idea in a general way, and a similar argument was employed by Kant (see Henry Allison, "Kant's Refutation of Materialism," *Monist* 79 [April 1989]). The argument was employed by C. S. Lewis, and was criticized by Elizabeth Anscombe in a famous confrontation at Oxford's Socratic Club; for an insightful commentary see Reppert, "Lewis-Anscombe Controversy." Other recent treatments include Warner Wick, "Truth's Debt to Freedom," *Mind*, n.s., 73 (1964): 527–37; James N. Jordan, "Determinism's Dilemma," *Review of Metaphysics* 24 (1969–70): 48–66; and Karl Popper, "Of Clouds and Clocks," in *Objective Knowledge: An Evolutionary Approach* (Oxford: Clarendon, 1972). An earlier treatment of my own is "The Transcendental Refutation of Determinism," *Southern Journal of Philosophy* 11 (1973): 175–83.

customed helping of humble pie by adopting a thoroughly evolutionary epistemology in which our theories take our place in the struggle for survival, some of them dying in our place in order that we might live.[33] Other recent philosophers who look to evolution to guarantee the general soundness of our cognitive strategies include Quine, Dennett, Fodor, Goldman, Lycan, Millikan, and Papineau.[34]

One dissenter from this chorus of optimism is Stephen Stich, who contends that "there are major problems to be overcome by those who think that evolutionary considerations impose interesting limits on irrationality."[35] Evolution, he argues, may well have left us with cognitive systems that are far from ideally rational—and there is significant empirical evidence that this is in fact what has happened.[36]

The truth is far, far worse than even Stich has imagined. If we accept the physicalist premises of causal closure and the supervenience of the mental, Darwinist epistemology flunks out completely: it has no ability whatever to explain how any of our conscious mental states have even the most tenuous hold on objective reality. Let me explain why this is so by sketching out very briefly the central ideas of evolutionary epistemology, to be followed by an equally sketchy outline of Darwinian evolution viewed in the perspective of physicalism. By juxtaposing the two, we will be able to see the fatal flaw in the combined project.

The central idea of Darwinist epistemology is simply that an organism's conscious states confer a benefit in the struggle to survive and reproduce. Such responses as discomfort in the presence of a chemical irritant, or the awareness of light or warmth or food, enhance the organism's ability to respond in optimal fashion. For more complex animals there is the awareness of the presence of predator or of prey and the ability to devise simple strategies so as to increase the chances of successful predation or of escape therefrom. As the organisms and their brains become more com-

---

33. See especially Karl R. Popper, *Objective Knowledge: An Evolutionary Approach* (Oxford: Clarendon, 1972). For the humble pie, see p. 241.

34. For references, see Stephen Stich, "Evolution and Rationality," in *The Fragmentation of Reason: Preface to a Pragmatic Theory of Cognitive Evaluation* (Cambridge: MIT Press, 1990).

35. Ibid., p. 56.

36. See Stich, "Could Man Be an Irrational Animal? Some Notes on the Epistemology of Irrationality," in Hilary Kornblith, ed. *Naturalizing Epistemology* (Cambridge: MIT Press, 1985), pp. 249–67.

plex, we see the emergence of systems of beliefs and of strategies for acquiring beliefs. Natural selection guarantees a high level of fitness, including cognitive fitness:

> Inferential strategies that generally yield true beliefs are fitness enhancing, and . . . natural selection will favor them. This, it is urged, is because in general having true beliefs is more adaptive than having false ones. True beliefs enable an organism to cope better with its environment; they enable the organism to find food, shelter, and mates, to avoid danger, and thus to survive and reproduce more effectively. There are exceptions, of course. . . . All the advocate of the current argument need claim is that on the whole and in the long run, organisms will be more fit—they will outcompete their conspecifics—if their ratio of true beliefs to false ones is higher. If this is right, then we can expect that natural selection will prefer one inferential system to a second if the former does a better job of producing truths and avoiding falsehoods.[37]

Our summary of evolutionary theory can be even more austere. It goes like this: Certain complex assemblages of organic chemicals develop a kind of dynamic stability in their interactions with the environment, together with a capacity for self-replication, which leads us to say they are alive. A variety of random physical forces leads to variations in the self-replicating assemblages, and some of the assemblages are more successful than others in maintaining and reproducing themselves. Over time, some of these assemblages become more complex than the earliest forms by many orders of magnitude, and their behaviors and interactions with the surrounding environment also become more complex. Nevertheless, the entire process is governed by, and explicable in terms of, the ordinary laws of physics and chemistry. Put differently, it is never necessary to go outside of the physical configurations and the physical laws in order to predict the future behavior of these assemblages; this is the "closure of the physical."[38]

37. Stich, "Evolution and Rationality," pp. 38–39. Stich is a skeptic about this argument, but he provides this formulation as a basis for discussion.

38. This summary is taken from my article, "Theism and Evolutionary Biology," in Philip Quinn and Charles Taliaferro, eds., *A Companion to the Philosophy of Religion* (Oxford: Blackwell, 1997), pp. 430–31.

It is hoped that the reader will already have spotted the incongruity of these two accounts. It's not merely that the Darwinist account doesn't *mention* the adaptive benefits of awareness and cognition. If that were the only problem, it could easily be maintained that awareness and cognition are among the necessary preconditions for the more successful behaviors and interactions with the environment that are featured in the account. (That, in fact, is exactly what evolutionary epistemology affirms to be the case.) The problem, rather, is that the Darwinist account *precludes* the kind of role for awareness and cognition that is posited in the epistemological account. It does this by its last two sentences, which affirm the causal closure of the physical domain. Those sentences guarantee that the conscious state of the organism, as such, can have *no influence whatever* on the organism's behavior and thus on its propensity to survive. The central contention of evolutionary epistemology has been decisively undermined.

To this it will be replied that the organism's conscious states are supervenient upon, and perhaps even identical with, its brain-states, and as such they do after all have a causal influence on behavior. This brings us back to considerations discussed at length in the previous section. It is of course true that if mental events are identical with physical events, they will have causal influence. For the moment, I'll concede that this is so even if they are not identical with, but merely supervenient upon, physical events. But as Kim has said, "whether or not a given event has a mental description . . . seems entirely irrelevant to what causal relations it enters into. Its causal powers are wholly determined by the physical description or characteristic that holds for it; for it is under its physical description that it may be subsumed under a causal law." *The mental properties of the event are irrelevant to its causal influence.* One simply cannot say whether the organism's behavior would have been different had the action been unreasonable from its standpoint, though one can say with assurance that the behavior would not have occurred had its physical sufficient condition been lacking.

The previous analysis in terms of possible worlds applies here as well. If we assume that mental events are nomologically necessitated by physical events, then a world in which this did not happen, or in which wildly different mental events supervened on the same physical events, would be *indistinguishable from the actual world* with respect to the physical course of events. If we assume that mental events are metaphysically necessitated by physical events, then it will not be possible to have the actual world stripped of its mental components. But there is still the physically equivalent zombie-world, one indistinguishable from the actual world in all physically observable respects, but completely lacking in mentality. In such a world organisms would still flee from danger, seek food and sex, and complain about the weather (or, emit sounds interpretable as complaints about the weather), just as they do in the actual world. The lack of any actual subjective states would make *no difference whatever* to the physical course of events, or to the survival or perishing of any creature.

What this means is that, given the physicalist assumption, *the occurrence and content of conscious mental states such as belief and desire are irrelevant to behavior and are not subject to selection pressures.* On this assumption, *natural selection gives us no reason to assume that the experiential content of mental states corresponds in any way whatever to objective reality.* And since on the physicalist scenario Darwinist epistemology is the *only* available explanation for the reliability of our epistemic faculties, the conclusion to be drawn is that physicalism not only *has not given* any explanation for such reliability, but it *is in principle unable to give* any such explanation. And that, it seems to me, is about as devastating an objection to physicalism as anyone could hope to find.

Evidently, this argument is closely related to the argument in the preceding section. But there is this difference: That argument depended on some crucial epistemological assumptions—for example, about the nature and causal influence of rational insight. These assumptions could in principle be denied—though not, in my opinion, with a great deal of plausibility. But no such epistemological assumptions are involved in the present argument. The only assumptions it requires are, first, that the conscious mental states of humans are systematically in contact with or have a grasp upon reality, and, second, that this fact requires some sort of explanation. These assumptions are indeed minimal, and all non-eliminativist versions of physicalism are already committed to accepting them. The prospects for evading the argument are correspondingly bleak.

Once the pieces are assembled, the argument given here seems fairly obvious, and one wonders why it has been overlooked until now. I am not sure of the reason for this, but a possible explanation is the following: When they write about rational processes, physicalists typically deal with them cybernetically, using notions of "information," "representation," and the like that can be treated within the scheme of physical explanation.[39] We thus become accustomed to the idea that these notions are at

39. The work of Dretske provides a good example of this.

home within the physicalistic scheme—though to be sure, difficulties are encountered in the details of such accounts. If the physicalists in question are thoroughgoing functionalists, the issue of conscious experience is never addressed, so of course the question of the correspondence of experience to reality never emerges. If on the other hand consciousness is brought into the picture, it is taken for granted that the mental experience supervenes on the physical state in such a way that (for instance) the cybernetically defined state identified with the belief that there is a cow in the pasture is accompanied by the kind of conscious awareness we would normally associate with assenting to that belief. What the present argument brings out is that *the correspondence between subjective experience and objective reality is an enormously important fact which requires explanation.* But on the assumptions of physicalism, no explanation can be given—this correspondence, *which we all assume to exist,* has the appearance of sheer miracle.

What conclusion should be drawn from this argument? At a minimum, the causal closure of the physical, and with it the commitment to universal mechanistic explanation, have to be given up. *Conscious mental states have to be recognized as causally effective precisely in virtue of their mental content.* If this is admitted, these states can have an effect on behavior and can be subject to selection pressures, and Darwinist epistemology is back in business. This may not be enough; there are still all the doubts raised by Stich to be taken account of. But nothing less offers any hope at all.[40]

40. The argument of this section was suggested, in part, by Alvin Plantinga's argument for the irrationality of naturalism in *Warrant and Proper Function* (New York: Oxford University Press, 1993).

Free Will and Agency

E mpirically speaking, there is not much of a case for determinism. The only direct empirical evidence for determinism is the existence of consistent, reliable, and accurate predictions of individual events.[1] In some fields of science we do have such predictions to a remarkable degree, but in others they are conspicuously absent.[2] To be sure, the failure of prediction in many areas of science can be explained in ways that are consistent with an underlying determinism. Perhaps the causal factors involved are simply too complex for our analysis— or (as chaos theory has shown) long-range outcomes may depend with incredible sensitivity on minute (and undetectable) differences in initial conditions. And the indeterminism of quantum mechanics, even if it is ultimate and not merely apparent, can in many contexts be ignored as making no difference to the behavior of macroscopic objects.

So deterministic theories *can be* maintained in the absence of reliable, accurate, predictions—but why *should* they be maintained? Why not admit that determinism just does not apply to certain aspects of the

---

1. It should be noted that statistical predictions afford no evidence at all for determinism, though they are sometimes fallaciously adduced as such. As the theory of probability shows, it is often possible to make statistical predictions even if we assume that the individual events are entirely random in their occurrence.

2. The weakness of the scientific case for determinism is developed at length in John Dupre, *The Disorder of Things: Metaphysical Foundations of the Disunity of Science* (Cambridge: Harvard University Press, 1993).

world—for instance, to the behavior of living creatures, and in particular of human beings? The reasons for this preference—and it is a powerful preference in recent science and philosophy—lies not in the empirical facts but in certain metaphysical preconceptions. Chief among these, no doubt, is the deeply entrenched commitment to mechanistic materialism, but there are others that manifest themselves at various points in the discussion. Anyone hoping to gain a hearing for a serious indeterminism and/or for a libertarian conception of free will, needs to identify and contest these preconceptions.

This is our task in the present chapter. In the first section, we will sketch the outlines of a libertarian conception of free will, and will identify (but not pursue in detail) some of the considerations in its favor. Subsequent sections will investigate some of the principal objections to libertarianism and will attempt to set out a viable libertarian conception of free agency.

LIBERTARIAN FREE WILL

According to Thomas Nagel,

> Our ordinary conception of autonomy . . . presents itself initially as the belief that antecedent circumstances, including the condition of the agent, leave some of the things we will do undetermined: they are determined only by our choices, which are motivationally explicable but not themselves causally determined. Although many of the external and internal conditions of choice are inevitably fixed by the world and not under my control, some range of open possibilities is generally presented to me on an occasion of action—and when by acting I make one of those possibilities actual, the final explanation of this (once the background which defines the possibilities has been taken into account) is given by the intentional explanation of my action, which is comprehensible only through my point of view. My reason for doing it is the *whole* reason why it happened, and no further explanation is either necessary or possible.[3]

And according to John Searle,

3. Thomas Nagel, *The View from Nowhere* (New York: Oxford University Press, 1986), pp. 114–15.

If there is any fact of experience that we are all familiar with, it's the simple fact that our own choices, decisions, reasonings, and cogitations seem to make a difference to our actual behaviour. There are all sorts of experiences that we have in life where it seems just a fact of our experience that though we did one thing, we feel we know perfectly well that we could have done something else. We know we could have done something else, because we chose one thing for certain reasons. But we were aware that there were also reasons for choosing something else, and indeed, we might have acted on those reasons and chosen that something else. Another way to put this point is to say: it is just a plain empirical fact about our behaviour that it isn't predictable in the way that the behaviour of objects rolling down an inclined plane is predictable. And the reason it isn't predictable in that way is that we could often have done otherwise than we in fact did. Human freedom is just a fact of experience.[4]

What makes these descriptions of our experience especially striking is that neither Nagel nor Searle is a professed libertarian. Rather than their descriptions being biased by libertarian assumptions, each of them finds in his description an obstacle to accepting conclusions that are motivated by otherwise strong preconceptions. The result of this is that neither man thinks the problem of determinism and free will has received a satisfactory solution. Nagel, characteristically, says that his "present opinion is that nothing that might be a solution [to the problem of free will] has yet been described."[5] Searle, as best I can make out, is a determinist who nevertheless accepts that "for reasons I don't really understand, evolution has given us a form of experience of voluntary action where the experience of freedom, that is to say, the experience of the sense of alternative possibilities, is built into the very structure of conscious, voluntary, intentional human behaviour. For that reason, I believe, neither this discussion nor any other will ever convince us that our behaviour is unfree."[6]

Now the question posed by the libertarian is: Why shouldn't we take the descriptions offered by Nagel and Searle at face value, as accounts of the way the world really is? It's not as though determinism had such impressive credentials of its own. As we've already noted, the empirical case

4. John Searle, *Minds, Brains and Science* (Cambridge: Harvard University Press, 1984), pp. 87–88.
5. Nagel, *View from Nowhere*, p. 112.
6. Searle, *Minds, Brains and Science*, p. 98.

for determinism is flimsy at best. And which variety of determinism should we espouse? In the view of many philosophers, the most plausible form of determinism is the physicalistic version discussed in the previous two chapters. But if the arguments developed in the last chapter have any merit, this version of determinism faces grave difficulties. It isn't easy to make out a rational case for accepting a theory which, if correct, would imply that one can't make out a rational case for anything at all!

About the only plausible alternative is psychological determinism,[7] often described as determination "by the strongest motive." But as has often been argued, this formula is empirically vacuous, since the "strongest motive" can only be identified retrospectively, by seeing which motive has in fact led to action. If on the contrary we try to be seriously empirical, making testable predictions and checking the outcomes, the results so far give little reason for optimism. (As has already been pointed out, *statistical* predictions provide no good grounds for determinism. Successful statistical predictions are perfectly consistent with libertarianism, provided we accept that motives and circumstances have some effect on behavior.) A good many philosophers (Searle, for instance—and consider Davidson's "anomalism of the mental") concede that there is no good case for psychological determinism.[8] So the friends of determinism may not find it easy to pick a plausible version to defend.

For myself, I have to say that I find the sorts of descriptions given by Nagel and Searle extremely compelling. These descriptions give us a perspective which seems to be *internal* to the experience of acting and making decisions; it may therefore be simply impossible for us to avoid relying on it in practice, whatever our theoretical qualms about it may be. Rejecting this understanding of experience ought to be recognized as a major form of skepticism, along with skepticism about the external world, skepticism about other minds, and other varieties. When con-

---

7. Actually, theological determinism is a serious contender, but it won't appeal to very many of the scientific determinists whose views are being addressed in this chapter. In any case, many theological determinists also accept psychological determinism.

8. Searle writes, "Insofar as psychological determinism is an empirical hypothesis like any other, then the evidence we presently have available to us suggests it is false. Thus this does give us a modified form of compatibilism. It gives us the view that psychological libertarianism is compatible with physical determinism." He admits, however, that "this form of compatibilism still does not give us anything like the resolution of the conflict between freedom and determinism that our urge to radical libertarianism really demands" (*Minds, Brains and Science*, pp. 97, 98).

fronted with such skepticisms we always ought to ask ourselves whether the arguments in their favor, even if apparently cogent, are really of sufficient force to outweigh what seem to be palpable facts of experience. I don't want to claim that the experience of freedom described by Nagel and Searle *proves* that we really have libertarian free will, but it does establish a powerful presumption in its favor—a presumption that ought to be overcome only by the strongest possible reasons for the contrary position.

It seems to me, also, that the sense of autonomy described by these philosophers—the sense that we are, in a real sense, the architects of our own lives—is an important component of the intrinsic worth and dignity that many of us want to ascribe to ourselves and to other human beings. And the often-stressed connection between the "power to choose otherwise" and moral responsibility also seems compelling. How in reason can a person be held responsible—whether for good or for ill—for doing what she was ineluctably determined to do by forces that were in place long before she was born?

But these are, to say the least, well-plowed fields, and little would be gained by rehearsing all the arguments here. Instead, the present chapter will take the libertarian view as a datum and will proceed to examine objections and to resolve difficulties with this conception. In order to do this, however, it will be helpful to have a somewhat more formal statement of the libertarian view. A definition I have used in another context is as follows:

> N is free at t with respect to performing $A =_{df}$ It is in N's power at t to perform A, and it is in N's power at t to refrain from performing A. . . . In general, if it is in N's power at t to perform A, then there is nothing in the circumstances that obtain at t which *prevents or precludes* N's performing A at t. Here "prevent" applies especially to circumstances that are *causally* incompatible with N's performing A at t, and "preclude" to circumstances that are *logically* incompatible with N's doing so. (The tied hands *prevent* Thomas from performing on the parallel bars; he is *precluded* from marrying Edwina at t by the fact that at that time she is already married to someone else.)[9]

9. William Hasker, *God, Time, and Knowledge* (Ithaca: Cornell University Press, 1989), pp. 66–67.

There are a couple of features of this definition which, in the present context, merit special comment. First, notice the *two-way power* ascribed to N: N has the power to perform the action, and also the power to refrain from performing it. This represents the "alternative possibilities" that are crucial to standard definitions of libertarianism—and that are featured extensively in the descriptions given by Nagel and Searle. But second, notice the references in the definition to N's *performing the action A*. This notion is of course crucial not only for libertarianism but for action theory in general. But libertarian theories generally have taken this notion as a *primitive and irreducible*; it cannot be analyzed or explicated in terms of the behavior of impersonal or subpersonal entities. And this gives rise to the idea of "agent causation," which is perhaps the most challenging philosophical problem of all those that surround libertarian free will.

These two features define the agenda for the remainder of the chapter. In the next section, we will address an important challenge to the idea that alternative possibilities are required for free will and moral responsibility. The following section will examine an attempt to explicate libertarian free will without agent causation, and the final section will attempt to set out a viable doctrine of agent causation.

THE FRANKFURT COUNTEREXAMPLES

In 1969 Harry Frankfurt launched an assault, still claimed by many to have been successful, on the "principle of alternate possibilities," which he formulates as follows: "A person is morally responsible for what he has done only if he could have done otherwise."[10] He argues for this by providing a counterexample to the principle; once understood, the counterexample allows for the creation of numerous others with subtle variations.

It should be said at the outset that we will be considering Frankfurt's example from the standpoint of a *libertarian* conception of free will; we will understand it as attempting to show that the principle must be abandoned *even if we accept other assumptions that would be made by a libertar-*

---

10. Harry G. Frankfurt, "Alternate Possibilities and Moral Responsibility," in John Martin Fischer, ed., *Moral Responsibility* (Ithaca: Cornell University Press, 1986), p. 143; originally published in *Journal of Philosophy* 66 (December 1969): 828–39.

ian. Since libertarian definitions of free will (such as the one proffered above) commonly embody the principle, its abandonment would force either a redefinition of libertarianism or an outright capitulation to compatibilism. It should be noted, however, that a version of the principle may be accepted even by compatibilists,[11] and so Frankfurt's argument could be employed to show those compatibilists the error of their ways. In what follows we will take no account of this use of the argument.

Here, then, is Frankfurt's example:

> Suppose someone—Black, let us say—wants Jones[12] to perform a certain action. Black is prepared to go to considerable lengths to get his way, but he prefers to avoid showing his hand unnecessarily. So he waits until Jones is about to make up his mind what to do, and he does nothing unless it is clear to him (Black is an excellent judge of such things) that Jones is going to decide to do something *other* than what he wants him to do. If it does become clear that Jones is going to decide to do something else, Black takes effective steps to ensure that Jones decides to do, and that he does do, what he wants him to do. Whatever Jones's initial preferences, then, Black will have his way.[13]

Frankfurt goes on to offer the reader her choice of the means by which Black will ensure that Jones acts as Black wishes; these may include threats, hypnosis, or the direct manipulation of Jones's brain processes.

It's clear what Frankfurt's point is: If Jones does what Black wants on his own, without Black's having done anything in the situation, we should surely want to say Jones is responsible for his own action, in spite of the fact that, given Black's conditional intention to interfere, it was not in Jones's power to refrain from doing as Black desired. So the principle of alternative possibilities is false. The argument seems cogent, and even

11. Presumably the "could" of "could not do otherwise" would be interpreted by the compatibilist in a hypothetical sense, along the lines of "could . . . if she had so desired." Frankfurt states, "I do not propose to consider in what sense the concept of 'could have done otherwise' figures in the principle of alternate possibilities" (ibid., p. 148).

12. Frankfurt identifies the individual in question as "Jones$_4$," thus allowing for earlier examples in which other versions of Jones figure. For present purposes the subscript will be omitted.

13. Ibid., pp. 148–49.

confirmed libertarians sometimes concede that Frankfurt has refuted the principle. If they wish to remain libertarians, they will propose alternative principles that are invulnerable to the Frankfurt counterexamples.[14] I will argue, on the contrary, that the example *fails entirely* to refute the principle of alternative possibilities, as this principle is understood by a libertarian.

In order to see this, we need to make some careful distinctions. First, consider the actual process of making a decision. Making a decision typically takes some time, even a lengthy period, but at some point the process arrives at what may be termed an *effective intention*. Such an intention is a state of mind[15] which, in the normal course of events, flows naturally into the intended action; no further deliberation is required. In the simplest cases involving bodily movements, the effective intention may produce the bodily action by a process that is effectively deterministic; the agent no longer has control over the occurrence of the action, and, barring external interference, the action cannot but take place. Other cases require the agent's involvement over a period of time, and it may be possible for her to reconsider and change her mind—but barring this, the effective intention produces the action without any further decision being required. The moment at which the effective intention becomes actual is, properly speaking, the moment at which the decision is made.

The notion of an effective intention allows us to raise an important question about Frankfurt's example. *Precisely when is it that Black intends to intervene?* Will he do this *before* the formation of an effective intention by Jones, or *after* this has occurred? The answers to this give us two different types of cases, which can then be further subdivided according to whether the intervention by Black actually occurs or not.

Case I: The intervention, if it occurs at all, occurs *before* an effective intention has been formed. This is what is most naturally suggested by Frankfurt's language: Black "waits until Jones is about to make up his mind what to do, and he does nothing unless it is clear to him . . . that Jones is going to decide to do something *other* than what he wants him to do." And suppose that *(a)* Black does foresee that Jones will make the

14. For example, see Peter van Inwagen, "Ability and Responsibility," *Philosophical Review* 87 (April 1978): 201–24; reprinted in Fischer, *Moral Responsibility*, pp. 153–73.

15. The use of mentalistic language here is unavoidable, but dualism is not being assumed. For purposes of the present discussion, it is possible to assume that each mental state or event is identical with a brain-state or brain-event.

"wrong" decision, and intervenes. Then it's clear that Jones does not make any free decision at all, and he is not responsible for whatever happens as a result. But what about the case in which *(b)* Black foresees that Jones will decide as Black wishes, and so does nothing? Well, consider the situation just *after* Black has decided not to intervene, but *before* Jones has arrived at an effective intention. At this point, it *still remains open* what Jones will in fact decide. Black has concluded that Jones will do as Black wishes, and he is likely right; let's admit that Black is a shrewd judge of such matters. But Black isn't infallible,[16] and last-minute reversals can (and sometimes do) occur—so at the moment we're focusing on, Jones's decision about what to do *still has to be made.* He may (and probably will) decide as Black wishes. But it's also possible that he will make the opposite decision, and if he does so, then *for all that we have been told, he is perfectly able to execute that decision.*[17] So in this case Jones surely is responsible, but the principle of alternative possibilities is satisfied.

Case II: Let us suppose, on the contrary, that Black will intervene only *after* Jones has arrived at an effective intention. (We'll assume that Black is able to determine this accurately.) Now if *(a)* Jones has made the "wrong" decision, and Black does intervene, it's clear that Jones is not responsible for the outcome, since it was Black's interference, and not Jones's own decision, that brought it about. But what if *(b)* Jones's decision is agreeable to Black, and he does what Black wishes without any need for Black to intervene? Here we want to say that Jones is responsible, in spite of the fact that, given Black's intention to intervene, it would not have been possible for Jones to refrain from acting as he did. So here, for the first time, it looks as though we may have a plausible counterexample.

But consider the following truism: *Normally, actions take time to execute and are subject to the possibility of interference.* The point is utterly obvious, yet it has been strangely neglected. And this point carries with it an implication about the correct interpretation of statements ascribing powers to individuals: *Statements ascribing powers to act to finite persons must normally be taken to include the implicit qualifier: "provided there is no interfer-*

16. I refrain from considering here cases involving an *infallible* predictor; readers interested in such cases should consult Hasker, *God, Time, and Knowledge*, chaps. 4–7.

17. To be sure, Black might have a backup plan for intervention at a later stage, should his initial prediction prove to have been wrong. This would move the example into Case II.

ence."[18] And from this we may further conclude that *a person's power to perform an action is not negated by the fact that such an action might be subject to interference.*

But if we apply these conclusions to our Case IIb, we obtain the following result: At the time Jones made his decision (which accorded with Black's designs), he *did after all* have the open alternative of acting differently, and so the principle of alternative possibilities is satisfied! To be sure, if Jones had actually made that other decision, Black might have intervened and prevented him from carrying it out. But this fact no more takes away Jones's power to act differently than it takes away my present power to drive to the grocery store, that it is possible that someone might run into my car and prevent me from arriving there. So, contrary to Frankfurt's claim, we still have no counterexample to the principle of alternative possibilities.

But what if it is *causally impossible* for Jones to act contrary to Black's wishes? (Call this Case IIc.) Perhaps Black has set up a mechanical device which will somehow register Jones's contrary decision and intervene automatically, preventing Jones's action with no further volition on Black's part. We do not want to ascribe to anyone the power to do something that is causally impossible, so in this case it is finally true that Jones lacks the power to act contrary to Black's wishes. But while a person cannot have the power to do something that is causally impossible, a person may have the power to *decide* to do such a thing—to form an effective intention so to act—*provided he does not know that the act in question is impossible.* (I can decide to leave a room whose only door is locked from the outside, provided I do not know it is locked.) So Jones does after all have the power to *decide* to act against Black's wishes; indeed in the IIa version of the example he actually does so. Note, furthermore, than on any reasonable libertarian theory it is the *decision to act* to which moral praise or blame is ultimately attached, rather than to the overt action, which may (as in this case) be frustrated by factors for which the agent is in no way responsible. So in the way that matters for moral responsibility Jones still has, even in this case, an alternative to doing what Black wants him to do. The principle of alternative possibilities emerges unscathed.[19]

18. There would be an exception for actions that are essentially instantaneous, and so cannot be prevented once the effective intention is formed.

19. A defense of libertarianism somewhat similar to the one developed here is given by David Widerker, "Libertarian Freedom and the Avoidability of Decisions," *Faith and Philosophy* 12 (January 1995): 113–18.

John Martin Fischer takes note of the kind of response just given to Case IIc; it is one version of what he terms (not inappropriately) the "flicker of freedom strategy."[20] His response is that although this strategy does succeed in identifying an alternative possibility, "this alternative possibility is not sufficiently *robust* to ground the relevant attributions of moral responsibility."[21] He says,

> The existence of *various* genuinely open pathways [into the future] is alleged to be *crucial* to the idea that one has *control* of the relevant kind. But if this is so, I suggest that it would be very puzzling and unnatural to suppose that it is the existence of various alternative pathways along which one does *not* act freely that shows that one has control of the kind in question. How exactly *could* the existence of various alternative pathways along which the agent does *not* act freely render it true that the agent has the relevant kind of control (regulative control)?[22]

Unfortunately, Fischer fails here to take seriously the beliefs of the libertarians he is seeking to refute. Libertarians do *not* consider that responsibility is grounded in the existence of "alternative pathways along which the agent does *not* act freely." For them, responsibility is grounded fundamentally in one's *freely* forming an effective intention (or, as it is sometimes called, a volition) to act in a certain way. In the normal case the intention, once formed, is carried out in an overt action. But if this for some reason does not happen (and this is the heart of the Case II Frankfurt examples), it is missing the point entirely to dismiss the alternative possibility as "not robust" and "not the right sort" to ground responsibility. On the contrary, to make responsibility fundamentally dependent on the overt act is to make it dependent on something over which the agent ultimately lacks control and for which he *cannot* be responsible. This is precisely the point of Kant's praise of the good will, which, if frustrated by circumstances, nevertheless "would sparkle like a jewel in its own right, as something that had its full worth in itself."[23]

20. See John Martin Fischer, *The Metaphysics of Free Will: An Essay on Control* (Oxford: Blackwell, 1994), pp. 134–47.

21. Ibid., p. 140.

22. Ibid., p. 141.

23. Immanuel Kant, *Foundations of the Metaphysics of Morals*, trans. Lewis White Beck (Indianapolis: Bobbs-Merrill, 1959), p. 10.

It may be that the complicated circumstances of the example tend to cause confusion about what would be entirely clear in a simpler situation. Suppose a young girl, told not to leave the house while her mother is at the store, obediently spends the time doing her homework. Surely she deserves full credit for her obedience even though, unknown to her, her mother has locked all the doors so that it would be impossible for her to leave. Or consider the case of someone who willfully refuses potentially life-saving medication to a dying patient. Such a person is certainly guilty even though, unknown to all concerned, the illness has been misdiagnosed and the medication could not have done the patient any good.[24] Does Fischer really think that in these cases the alternative possibilities are not sufficiently robust to ground moral responsibility?

The defender of Frankfurt may have yet one more arrow in his quiver.[25] According to some philosophers, there exists for each free person a set of truths stating what that person would *freely* (in the libertarian sense) choose to do in various possible circumstances. It has become customary to refer to these truths as "counterfactuals of freedom," and they play a role in certain controversies concerning the nature of divine knowledge.[26] Now, suppose that there are such truths about Jones, and that Black is somehow in a position to know them. Given this, we can set up *Case III*, in which, rather than judging from perceptual clues what Jones is going to decide (as in Case I), Black consults the counterfactuals of freedom concerning Jones, and learns from these whether, under the existing circumstances, Jones would choose in accordance with Black's wishes or against them. In the latter case (IIIa), Black intervenes and, as in Case Ia, Jones makes no free choice for which he could be responsible. But Case IIIb, in which Jones freely makes the "right" choice, seems to represent a last hope for Frankfurt. For once again, with no interference from Black, Jones is surely responsible for his own decision. But it seems there is no alternative possibility, for if Jones had been going to choose to act otherwise, he would not have been able to do so.

24. Here we must be careful to distinguish moral responsibility from legal liability; the latter, much more than the former, depends on actual harm being done.

25. John Martin Fischer raises the possibility of a Molinist defense of Frankfurt in "Libertarianism and Avoidability: A Reply to Widerker," *Faith and Philosophy* 12 (January 1995): 119–25. It was Luis de Molina, a seventeenth-century Jesuit, who first proposed the doctrine of divine "middle knowledge," which employs the "counterfactuals of freedom."

26. For details, see my *God, Time, and Knowledge*, chap. 2.

This however is a confusion. It is true that *if Jones had been going to choose to go against Black's wishes, he would not have been able to do so.* What is *not* true, however, is that *if Jones had chosen to go against Black's wishes, he would not have been able to do so.*[27] Remember that the counterfactuals known by Black are counterfactuals of *freedom*; they describe what Jones would *freely* choose to do in various circumstances. If Black does not intervene, then Jones will indeed freely choose what he is going to do. And if he were to choose against Black's wishes, no reason has been given why he would not be able to carry out his choice in action. To be sure, we know (and Black knows) from Jones's counterfactuals of freedom that Jones *will not in fact* choose that way. But his doing so is neither logically impossible, nor causally impossible, nor beyond Jones's power. So there are after all alternative possibilities, and the counterexample fails.

Now there may be something odd about the notion that there is a genuinely free choice with real alternative possibilities, and yet it is possible for someone to know with certainty how the choice would in fact be made. If, like me, you find this implausible, you may also want to join me in rejecting the existence of true counterfactuals of freedom. (Or, you might concede their existence but deny that they are knowable.) But this problem, if it is a problem, concerns the nature, truth, and knowability of those very peculiar propositions; it is not a problem about alternative possibilities.

I conclude, then, that the Frankfurt counterexamples are vastly overrated; philosophers who rely on them to refute the principle of alternative

27. The formal logic here is a bit complex. Let

C = Certain circumstances obtain (those that will obtain in the actual world, if Black does not intervene first).

D = Jones decides to act in a way contrary to that desired by Black.

P = Black prevents Jones from acting contrary to Black's wishes.

Now, one might reason as follows: If both C and D are true, this entails that the counterfactual (C $\square\!\!\rightarrow$ D) is true. (This is implied by the theory of middle knowledge.) If (C $\square\!\!\rightarrow$ D) were true, then Black would prevent Jones's action (i.e., P would be true). So, if C and D were both true, then Black would prevent Jones's action. In symbols:

1. (C & D) entails (C $\square\!\!\rightarrow$ D)
2. (C $\square\!\!\rightarrow$ D) $\square\!\!\rightarrow$ P

Therefore, 3. (C & D) $\square\!\!\rightarrow$ P

This reasoning, however, is fallacious, because the inference rule "P entails Q; Q $\square\!\!\rightarrow$ R; therefore, P $\square\!\!\rightarrow$ R" is invalid.

possibilities—and to defeat libertarianism—need to go in search of better arguments.

FREE WILL WITHOUT AN AGENT

I will proceed, then, on the assumption that we do have free will in the libertarian sense, and that the principle of alternative possibilities is intact. Our concern in this section and the next will be with the other, perhaps even more difficult, part of the free will problem: the task of giving an illuminating *positive characterization* of the nature of the act of free choice. For the libertarian will always have to confront the compatibilist's contention that *chance* and *randomness* are even more inimical to freedom and responsibility than is causal determination. As Hume explained (and his argument has been echoed ever since): "Actions are, by their very nature, temporary and perishing; and where they proceed not from some *cause* in the character and disposition of the person who performed them, they can neither redound to his honour, if good; nor infamy, if evil."[28] The challenge for the libertarian is to explain how free actions are praiseworthy or blameworthy even though they lack a sufficient cause, whether in the agent's character or elsewhere.

In approaching this problem, we shall look for help to an unexpected source, namely Daniel Dennett's essay "On Giving Libertarians What They Say They Want."[29] In this essay Dennett sketches out, though he does not fully endorse, a strategy for explaining libertarian free will. Following a suggestion from David Wiggins, he undertakes to show that an action might be "random" in the sense of *causally undetermined*, without being random in the sense of *pointless* or *arbitrary*. (That is in effect what Hume had charged.) That is to say, he undertakes to show how an action might be causally undetermined, yet *intelligible* (and, in some cases, even *predictable*) in terms of reasons. In his first example, a subject has to answer a yes or no question. The subject knows the correct answer (which is Yes), and is highly motivated, by a promised reward, to give that answer. However, the physical movement the subject will make in answering the question is undetermined, for the following reason: The correct (Yes) an-

---

28. David Hume, *An Enquiry Concerning Human Understanding* (La Salle, Ill.: Open Court, 1963), p. 107.

29. In Daniel C. Dennett, *Brainstorms* (Cambridge: Bradford Books, 1978), pp. 286–99.

swer is to be signaled either by pushing a button with the right hand or by stepping on a pedal with the right foot. At any given time, *either* the button or the foot pedal is activated, but not both. Which is activated at the time when the question is asked is determined by a randomizing device, which, operating in a causally undetermined fashion, switches the pedals "on" and the buttons "off" (or vice versa) and also signals to the subject which he is to use to record his answer. In this case one can *predict*, on the basis of his reasons, what the subject will do (he will answer Yes), but the *physical movement* by which he will signal his answer is in principle unpredictable, because it will be affected by the causally undetermined randomizer. So we have a movement which is rationally explicable, even predictable, yet is causally undetermined.[30]

Now this is a far cry from anything a libertarian would recognize as free will—indeed, Dennett himself describes the example as a "cheap trick."[31] Nevertheless, it prepares the way for the second example. Here, the randomizer is moved to within the subject's brain. The subject has decided, for excellent reasons, to purchase a can of Campbell's tomato soup in the supermarket. There are a number of cans which are available, within easy reach, and with nothing to make one of them preferable to another. At this point, the randomizer kicks in, and directs the subject's hand and arm to pick up one can rather than another. Once again we have an action which on the level of reasons is entirely predictable—the subject chose a can of tomato soup—but the *physical movements* are undetermined and in principle unpredictable because of the presence of the randomizing device.[32]

This is perhaps a bit closer to free will than the previous example, yet it is obviously unsatisfactory in that the difference made by the randomizer is utterly trivial. What we need is a case in which what is causally undetermined is not merely *how* an action is done, but *what* is done. Enter the third example, in which a fortunate young academic is deciding between a job offer from the University of Chicago and one from Swarthmore. This decision, unlike the one between cans of soup, requires some deliberation. The subject will cast around, in the time available to her, for relevant considerations which tell in favor of one offer or the other. "Now it just might be the case," says Dennett, "that *exactly* which considerations occur to one in such circumstances is to some degree strictly undetermined."[33] As it

30. Ibid., pp. 289–90.
31. Ibid., p. 289.
32. Ibid., pp. 290–91.
33. Ibid., p. 294.

happens, the randomizer brings up to her mind considerations A, B, C, D, E, and F, and on this basis she chooses Swarthmore. It was causally possible, however, for the randomizer to have brought G to her attention, and if it had done so she would have chosen Chicago. So her action was not *predictable*, even for one with a thorough knowledge of her previous character, preferences, etc. Nevertheless, the choice of Swarthmore is thoroughly *understandable*, and on the other hand the other decision (had she thought of G and opted for Chicago) would also have been understandable. And here, finally, we have Dennett's proposed model for libertarian free will. He sums up his proposal as follows:

> The model of decision making I am proposing has the following feature: when we are faced with an important decision, a consideration-generator whose output is to some degree undetermined produces a series of considerations, some of which may of course be immediately rejected as irrelevant by the agent . . . Those considerations that are selected by the agent as having a more than negligible bearing on the decision then figure in a reasoning process, and if the agent is in the main reasonable, those considerations ultimately serve as predictors and explicators of the agent's final decision.[34]

As we've noted, Dennett does not fully endorse this model,[35] but he does feel it has impressive merits.[36] Rather than linger over these, however, I wish to call attention to a striking fact: Dennett's model for free will coincides, in its most important features, with that proposed in one of the best and most insightful books of recent years advocating libertarian free will: I am referring to Robert Kane's *Free Will and Values*.[37] In virtue of this similarity, I shall refer to their common doctrine as *DK-libertarianism*.

In noting this similarity, however, I may have raised a question in the reader's mind: Why, given the choice, did I expound Dennett's proposal here rather than the much more subtle and elaborate version of DK-liber-

34. Ibid.

35. He argues that, in order to obtain the benefits of the model, it is not really essential that the randomizer be *causally undetermined* in its operation; it would suffice that the randomizer be *patternless*. But if this change is made in the model, then free choices would after all be causally determined (ibid., p. 298).

36. For these see ibid., pp. 295–97.

37. Kane, *Free Will and Values* (Albany: SUNY Press, 1985). Kane acknowledges Dennett's proposal on pp. 103–4; Kane, however, had developed the idea independently before Dennett's essay appeared.

tarianism given by Kane? In part, the reason is a matter of expository convenience. Dennett's proposal, which he presents within a few pages, is much more easily summarized than Kane's development of the view, which is spread over several chapters. But I must confess to an ulterior reason as well. I am hoping, frankly, that the provenance of the model will motivate questions about its adequacy. I am suggesting that libertarians should ask themselves, "With Dennett for our friend, why do we need enemies?" Dennett's contributions to the philosophy of mind are numerous and impressive, but few of his doctrines are such as to be welcomed enthusiastically by libertarians—indeed, he has become known as an outstanding contemporary advocate of compatibilism.[38] Surely a proposal for understanding libertarianism which is acceptable (or nearly acceptable) to Dennett requires the most careful scrutiny?

One nontrivial difficulty for DK-libertarianism is that it requires one to postulate within the brain a process by which quantum-level indeterminacy is amplified to produce macroscopic effects.[39] While this notion has seemed attractive to many, the general trend of brain science seems to be unfavorable to it, though the issue is not yet closed. So this assumption clearly represents a significant empirical burden for DK-libertarianism to carry.

Quite apart from this, the theory is hardly without difficulties. Kane observes that some libertarians will find his theory inadequate because it does not permit "total rational control" for the free agent.[40] It is clear enough what he has in mind. Consider a situation in which one comes to make a decision in which the reasons on both sides are about equally balanced, so that neither set of reasons necessitates a particular outcome. The decision actually made, in this case, is the product of chance—of the internal "randomizing device" about which we learned from Dennett. Is the outcome then *merely* the result of chance? No, because the range of possibilities available to one—the fact that *these* choices, and no others, are the "live options" in the situation—is itself a consequence of the reasons one has, and thus of the underlying values which are a stable part of one's personality. But *why* are these particular reasons, and values, part of the choice situation? One could, given sufficient knowledge, trace the causal history of those values and reasons in one's brain and nervous system. But

38. Especially in Dennett, *Elbow Room* (Cambridge: MIT Press, 1984).

39. For a discussion of this see Kane, *Free Will and Values*, chap. 9; also, John Thorp, *Free Will: A Defense against Neurophysiological Determinism* (London: Routledge and Kegan Paul, 1980), pp. 67–71.

40. See, e.g., Kane, *Free Will and Values*, p. 156.

somewhere back along the line, one comes upon another chance event—either the chance involved in the fact that *these* reasons, and not others, occurred to one, or the chance determination of some previous actions which led to those particular values becoming reinforced as a part of one's personality. And so on. When we complain about randomness, we are given (partial) causal determination by reasons. When we ask about the origin of the reasons, we are referred back to prior chance events, occurring of course within the context of reasons which were efficacious at that previous time. What we have, then, is an alternation of causal necessitation and chance;[41] what we never get, is a *person* making a *decision* in what a libertarian will take to be the ordinary sense of those terms.[42]

41. Mark Bernstein writes: "Valerian Libertarianism [his term for DK-libertarianism], however, should not be acceptable to Libertarians. Perhaps the easiest way to see this is to realize that after the initial potential considerations are randomly provided, the deliberation process runs deterministically. But since the Libertarian's incompatibilist intuitions find Determinism an anathema to freedom, it is difficult to understand how an account which incorporates chance succeeded by determinism can provide him . . . with a satisfying theoretical backdrop in which to understand personal freedom" (Bernstein, review of *Free Will and Values*, by Robert Kane, *Noûs* 23 [September 1989]: 559).

Kane, however, has stated to me that "the association of my view with Dennett's suggestions in *Brainstorms* has to be qualified a good deal more than people have been willing to do. . . . I'm fully in agreement about the limitations of DK or Valerian libertarianism as your paper outlines them. But it is not my theory, only part of my theory. . . . DK-libertarianism is not even the most crucial part of my theory, because my accounts of prudential and moral choice don't employ it in the same way, and in practical choice, where it plays a role, the key element which makes all the difference is the 'effort of will' to keep one's mind open . . . which is added at the end of chapter 6 and in chapter 8" (personal communication). Kane also mentions his paper "Two Kinds of Incompatibilism," in Timothy O'Connor, ed., *Agents, Causes, and Events: Essays on Indeterminism and Free Will* (New York: Oxford University Press, 1995); originally published in *Philosophy and Phenomenological Research* 50 (December 1989); the reader is referred to this article, as well as to Kane's book, and also to his more recent book, *The Significance of Free Will* (New York: Oxford University Press, 1996).

My own view is that the basic criticisms given above remain in effect in spite of the subtle elaborations Kane has provided. But the matter deserves careful consideration, and some readers may find Kane's view to be a viable alternative to the theory of agent causation defended here.

42. In a way, Kane recognizes that his theory is less than wholly adequate: "The task for the libertarian is to grab hold of this deviant child (partial control) and drag it back as close *as possible* to the norms of common sense. One cannot get all the way back, but no view on the free will issue fully satisfies common sense" (*Free Will and Values*, p. 155).

I believe, in fact, that the strategy of DK-libertarianism at this point is basically a compatibilist strategy. The compatibilist recognizes that straightforward determination of actions by external causes negates freedom and responsibility. His answer to this is to display the causal chains as interwoven, in subtle and complex ways, with the inner life of the agent. Similarly, the DK-libertarian recognizes that straightforward determination *either* by sufficient causation *or* by chance is inimical to freedom and responsibility. And the response is similar: the causation *and* the randomness are shown as being interwoven in complex ways with the agent's inner life, with the chance events actually occurring within the agent's nervous system. I shall not debate at length the merits of these strategies. Rather, I shall content myself with pointing out that libertarians consistently reject such a strategy as followed by compatibilists, and if so they should reject it also when put forward by DK-libertarians.

But what then? If free will can be explicated neither by causal necessitation, nor by chance, nor by any combination of the two however subtle, then how are we to explicate this crucial notion?

AGENT CAUSATION

The traditional libertarian answer to questions about the nature of free choice is found in the theory of *agent causation*. Recently, however, this theory has been more often perceived as a liability than as a strength of the libertarian view. Typically it is characterized as obscure and mysterious, if not downright unintelligible. In what follows I will attempt to put forward a version of agent causation which cannot fairly be rejected on this basis. I believe much of the perceived obscurity results from the imposition on the theory of requirements that are alien to it and that presuppose the truth of metaphysical views a libertarian should reject. That is not to deny that there are genuine difficulties. At several points there are choices to be made, and other agent causalists may make them differently than I have. So my claim here is not to have presented the unique "correct" version of agent causation; rather, I put forward a version I take to be viable and free from crippling objections, in the hope that others will carry on the discussion.

It well be helpful in what follows if the reader will first think through and thereafter keep in mind an example of a person's making a choice in accordance with reasons one has—an example that can naturally be described in the sort of terms set out in the quotations from Nagel and Searle. Take, for instance, the simple fact that you are continuing to read the present discussion. No doubt there are other things you could well be doing instead! There may be exams waiting to be graded, or lectures needing to be prepared—and wouldn't this be a good time to check your e-mail? Nevertheless, you continue to read this. Perhaps you hope to refute it in a paper you are writing—or, it may even be that you are interested in the line of argument and want to see how it comes out. Almost certainly, there is no external power or force that compels you to continue reading, and most likely the reasons for continuing to read are not so irresistible that you couldn't very well decide to break it off. This is, in Searle's words, one of those "experiences that we have in life where it seems just a fact of our experience that though we did one thing, we feel we know perfectly well that we could have done something else."

I suggest, then, that you keep this example (or one of your own devising) firmly in mind during the ensuing discussion. As various questions are raised about agent causation, try to bring to mind the sorts of things you would naturally say, in everyday life, about your example—provided, of course, you were not being challenged by contrary philosophical assumptions. And when you are confronted with arguments purporting to show that agent causation is somehow incoherent, ask yourself seriously whether these arguments are so cogent as to warrant your giving up as unintelligible something that, absent special philosophical assumptions, seems entirely coherent and reasonable.

Since agent causation is after all a variety of causation, the viability of agent-causal theories will be affected by one's general view of causation. If one accepts a reductionist theory of causation in which causal relations supervene on a basic layer of noncausal facts, agent causation is going to seem anomalous or even unintelligible. I believe, therefore, that some version of a realist theory of causation is most consonant with the requirements of agent causation.[43] An example of such a theory is found in Harré and Madden's *Causal Powers*, which argues for the locus of such powers in

---

43. This point is made by Randolph Clarke, "Toward a Credible Agent-Causal Account of Free Will," in O'Connor, *Agents, Causes, and Events*, pp. 207–8; originally published in *Noûs* 27 (1993): 191–203.

the "powerful particular."[44] Given a theory of this sort, both agent causation and what is ordinarily termed event causation are instances of *causation by substances*, in which the particulars in question produce their effects by exercising their inherent causal powers. The difference between the two types of cases would be that, in "event causation," the individual in question automatically produces the effect in response to a "triggering event," whereas in agent causation the substance is a person who, confronted with a situation involving various motives, opportunities to act, and surrounding circumstances, may act in any of several ways or may decide to do nothing at all. There is still plenty of difference between the cases, but this schema makes it intelligible that both are instances of *causation*; that is, of a substance producing an effect in virtue of its inherent causal powers. On the other hand, the applicability of "causation" to both types of cases may well be questioned if one accepts (for example) a Humean regularity theory.

The central elements of agent causation are given already in the quotations from Nagel and Searle. A person finds herself in a situation where there is more than one thing that might be done; typically, she has reasons for selecting two or more of the (mutually incompatible) options that present themselves. She then *decides* which of those options to actualize; her decision is guided by her motivations and values, but is not determined by them, so that under exactly the same circumstances she might have chosen differently.

An important feature of agent-causal theories is that the free choice is *made by the person as an integral whole*; it cannot be analyzed in terms of the behavior of impersonal or subpersonal constituents of the person. The reasons for this are implicit in discussions that have already been given. As regards the analysis of the person in terms of microphysical constituents, the arguments of the previous chapter are decisive: human action has to be understood as fundamentally teleological, but this is impossible if it is to be analyzed in terms of the physical interactions of particles which, in themselves, have no capacity to respond to normative or teleological considerations but instead simply follow the mechanistic laws that are natural to them. *The requirement that personal action should be analyzable this way is fundamentally alien to libertarianism.* If one nevertheless finds this re-

44. R. Harré and E. H. Madden, *Causal Powers: A Theory of Natural Necessity* (Oxford: Basil Blackwell, 1975); cf. Timothy O'Connor, "Agent Causation," in O'Connor, *Agents, Causes, and Events*, pp. 175–77.

quirement compelling—and is willing to pay the price of admitting that no one ever does anything because she has good reasons for doing so—then that is a reason (!) to reject libertarianism. But for the libertarian himself, who sees no reason to accept such a requirement, objections based on it are without force.

Another possibility, to be sure, is to analyze action in terms of such items as belief states, desire states, and the like. Here we do, presumably, have a responsiveness to normative and teleological considerations, and thus an account of action enormously more plausible than the account championed by the thoroughgoing mechanist.[45] So this kind of approach may be appealing to a compatibilist, but it fails as an account of libertarian free will, for reasons suggested in the previous section of this chapter. This approach captures the idea of an action being *done for reasons*, but the notion of a person's *choosing* the action, and thereby choosing *which* of her reasons will be the ones issuing in action, still can't be captured. The subject either has some "controlling reason" which determines the outcome, and so eliminates free choice, or else one reason dominates another through a chance mechanism—and once again, free choice is eliminated. Only the *self* or *person*, functioning as an integral whole, is able to exercise free will and make the choice.

But just how is this supposed to go? That is, what is actually going on when a person, confronted with alternative possibilities, chooses one of them in preference to another that she also has good reason (and perhaps equally good reason) to choose? A complete answer to this question would require an extensive phenomenology, something I am not prepared to provide. But a little reflection on some facts of experience should help us to see that we don't have here an insoluble conundrum. Let me point out, first of all, that it is often in our power to *direct our attention* in one way or another—to choose which of two things (for instance, of two desires we may have) we focus on, think about, and the like. And, secondly, that *the perceived strength of different motives often depends, to a substantial degree, on the attention that is given to them*. One of the clearest examples of this, perhaps, concerns the feeling of resentment for harms that have been done to us. The most elementary mental hygiene suggests that, even if wrong has genuinely been done to us, we are much better off not to be-

---

45. It needs to be stressed, though, that the two types of explanation can't possibly be combined, as many have tried, and still try, to do. That is, this combination is impossible *if* the scope of mechanistic explanation is assumed to be all-embracing.

come preoccupied with it, going over and over the offense until it fills our mental horizon. (Our current public obsession with "victimization" is questionable for precisely this reason.) And often (even if not always) it really is in our power, to a considerable extent, to control this. Some of us have been taught to pray, "Lead us not into temptation"; the prayer presupposes that one who sincerely prays this way will not unnecessarily *place himself* in a situation where the temptation to wrongdoing may become too appealing to be readily resisted. Examples could easily be multiplied, but I hope the point is clear: We do not usually find, in a choice situation, that different motives, desires, and so on present themselves as having predetermined "weights" which, in their interaction, produce a resultant "action vector" that inexorably carries us to one outcome rather than another. What we do find, in many cases, is a variety of motivating factors between which we have considerable ability to choose and thus to affect the outcome.

In such a situation, the action actually chosen can be given a "purposive explanation" in terms of the reasons the subject had for performing that action. Thus, one might explain one's continuing to read this chapter by saying, "What he's saying is really outrageous, and I'm looking for the fatal flaw in his argument." But of course, one could equally well explain making a different choice: "I meet my Intro class tomorrow, and I need to return their exams." If asked to explain why one of these reasons prevailed rather than the other, one might say, "It seemed more important to me to get the exams back; I can always finish the chapter later." It's important to see, though, that the "greater importance" attributed to returning the exams *results from* the way the choice was in fact made; it was not a "preexisting condition" which determined the outcome of the choice in advance. (Of course, it *can be* the case that, rationally considered, one set of reasons is clearly more weighty than another. And sometimes we choose the other way in spite of that. People *do* behave irrationally.)

All of this may seem, from a certain point of view, quite unsatisfactory. It's not enough, we will be told, that an action can be given a rational explanation after the fact. What we need is a *contrastive explanation* that explains, in terms of antecedent conditions, *why one alternative was chosen rather than the other*. And if the reply is "Well, that's the way I decided," or "At the time, that's what seemed most important," the challenge will simply be repeated: *Why* did you choose that way, rather than the other? *Why* did that seem most important at that time?

The proper response for a libertarian, of course, is simply to *reject* the demand for a contrastive explanation in such cases.[46] Such explanations are nice when we have them (they help us to make reliable predictions, for one thing), but the demand that they should be available in all cases is equivalent to asserting determinism as an a priori requirement on success-ful explanation. This requirement has no empirical basis, and if someone had a convincing a priori argument for it, it would still have to be squared with something we *do* have good empirical evidence for, namely the inde-terminism of quantum mechanics.

In order to further clarify the notion of agent causation, we pose and answer a series of questions:

What is the sense of "cause" in agent causation? The fundamental mean-ing of "cause" in the present context is production: the agent *produces* her own action. "Cause" *cannot* be understood in terms of regular succession, conformity to law, and the like, since it is postulated that there *are no* de-terministic laws connecting agents, circumstances, and actions. (And as we've noted, the postulate gains credibility from the failure of behavioral scientists to find such laws.)

If the agent causes her own action, does she cause her causing of her own ac-tion? The question makes no sense, if it implies a *simultaneous* causing of her causing her action. It makes sense only if what is being asked is whether some *previous* action of the agent's was in some sense a cause of her present action. This is so sometimes, but certainly not always.

This question arises, however, because of a verbal trap agent-causalists have not always avoided. One might ask, whether the agent-causing of an action is itself an action. To say that it is not may seem peculiar—surely, agent-causing is supposed to be something *done* by the agent, and if this isn't an action what is it? On the other hand, if agent-causing *is* an action then, according to the theory, it must itself be agent-caused, and now we have an infinite regress. Eminent agent-causalists—Reid,[47] for example, and also Chisholm[48]—have found themselves entangled in this kind of regress.

46. I find it surprising that Robert Kane, himself a libertarian, nevertheless presses this objection against agent causation; see his "Two Kinds of Incompatibil-ism," pp. 122–24.

47. See William L. Rowe, *Thomas Reid on Freedom and Morality* (Ithaca: Cornell University Press, 1991), pp. 147–51.

48. See Roderick M. Chisholm, *Person and Object: A Metaphysical Study* (La Salle, Ill.: Open Court, 1976), p. 71. Chisholm, however, later retracted the principles that lead to regress: see Chisholm, "Replies," in Radu J. Bogdan, ed., *Roderick M. Chisholm*, Profiles, vol. 7 (Dordrecht: D. Riedel, 1986), pp. 314ff.

It should be obvious, though, that this is *merely* a verbal trap. When I perform some action, such as lifting my hand, my agent-causing that action—that is, my forming an effective intention to lift my hand—is not a separate action distinct from lifting my hand. Rather, it is an essential component of the action of lifting my hand; it's what distinguishes the voluntary action, *my lifting my hand,* from *my hand's going up,* as might happen as a result of the electronic stimulation of a nerve. That the vital core, as it were, of an action should be something that is not itself an action may seem puzzling. But it is, to repeat, *merely* a verbal puzzle, and once we have realized that it should lose its power to bewilder us.

In causing the action, what exactly does the agent produce? The action as a whole is said to be caused by the agent, but the completion of the action may depend on matters over which the agent lacks control. What is immediately produced by the agent is, as we've said above, the "effective intention," or volition, a mental state which naturally, and without need for further deliberation and decision, leads to the action's being performed. (The effective intention thus intends to do something *here and now,* unlike an intention to do something at some time in the future.)

Can an agent-caused action be causally determined? Most agent causalists have said that it cannot. A notable exception, however, is Richard Taylor, who writes that "there is nothing in the concept of agency, as such, to entail that any events must be causally undetermined, and in that sense 'free,' in order for some of them to be the acts of agents. . . . What is entailed by this concept of agency, according to which men are the initiators of their own acts, is that for anything to count as an act there must be an essential reference to an agent as the cause of that act, whether he is, in the usual sense, caused to perform it or not."[49]

There are various possible responses to this. A common strategy is to build indeterminism into the definition of agent causation. But this invites the question: Could there not be instances that are exactly similar to agent-caused actions except that they are causally determined to occur? And if there could be, would these not be *actions*? Or, if they would after all be actions, would they be actions not caused by any agent?

In view of this possibility, a philosopher who wants to maintain that agent-caused actions cannot be causally determined would be well advised to search for some deeper, metaphysical reason for this rather than rely on

49. Richard Taylor, *Action and Purpose* (Englewood Cliffs, N.J.: Prentice Hall, 1966), pp. 114–15.

a stipulative definition. One philosopher who does this is Timothy O'Connor, who replies to Taylor as follows:

> Now it is one thing to argue in this way: it is perfectly intelligible that one should be determined on occasion to act as one does; on this theory, one is always the agent cause of one's acts; hence this theory is constrained to allow for the possibility that an agent is determined to cause his own action. But it is quite another directly to defend the idea of causally determined agent causation against the charge of incoherence. Just how are we to understand the notion of there being a sufficient causal condition for an exercise of active power?[50]

O'Connor goes on to argue that it is the metaphysical structure of the event, *S's causing e* (where *S* is the agent in question) that makes it impossible that this event should have a sufficient cause. He writes,

> Causally complex events can . . . be caused . . . but only in a derivative way: where they have the form *event X's causing event Y*, whatever causes event X is a cause *thereby* of *X's causing Y*. In the special case of an *agent's* causing an event internal to his action, however, there is no causally simple component event forming the *initial* segment, such that one might cause the complex event (*S's causing e*) in virtue of causing *it*. Therefore, it is problematic to suppose that there could *be* sufficient causal conditions for an agent-causal event.[51]

Consider, however, the following scenario: You are on a committee selecting a new member to be added to your department. The three finalists all have impressive credentials, and you are initially about evenly balanced between them. Upon further consideration, however, you realize that one of the candidates *far* outranks the others in terms of the qualities needed by the department. It strikes me as plausible to say that at this point you, as a conscientious person, are *unable* to cast your vote for one of the other candidates, or to refrain from voting for the best candidate, even though one of the others may be more appealing to you personally. Yet it is surely *not* plausible to deny that voting for the best candidate is an action you perform, and of which you are the agent-cause.

What O'Connor's discussion overlooks is that the event, *S's causing e*, es-

50. O'Connor, "Agent Causation," p. 185.
51. Ibid., p. 186.

sentially involves *S*'s *reasons* for causing *e*, and that on occasion (though certainly not always) one's reasons may be so compelling as to literally "leave one with no alternative." It's not a matter of having to alter the metaphysical structure of the event, *S's causing e*, in order to allow it to be causally determined. It's rather a matter of one of the ordinary constituents of the event, namely the reasons one has for performing an action, being so strong that they, in effect, preclude any other course of action.

In work currently in progress,[52] O'Connor considers this type of objection and makes a further response. He considers the case in which one has reason for only one action: "If having a reason to perform an action is necessary to freely forming the intention to do it, and one has, on occasion, reason to perform only one sort of action, is not that action thereby inevitable, even though the state of having the reason does not directly produce the action?" His reply, following Duns Scotus, is that one has at least the option of doing nothing. He then asks, "Does this only temporarily forestall matters? Well, either the ability to refrain from definite commitment persists, or the agent's state evolves in such a manner as to determine, at some point, the initiation of the act. In connection with the second possibility, it has been no part of my view that agents always act freely, only that when they do, their so acting was not then inevitable." Now, consider the case in which "the ability to refrain from definite commitment" does *not* persist. (The vote on the candidates must be cast; it can't be put off any longer.) Clearly O'Connor is right in saying that the agent in this case is not, in the libertarian sense, acting freely. But is the agent *not performing any act at all*? To say this seems extremely counterintuitive, and O'Connor's language strongly suggests that the agent *does* act in such a case. (If agents do not always act freely, then there must be times when they act unfreely.) But if the agent does act unfreely, it would seem that we have an agent-caused action that is determined to occur.[53]

Finally, *is agent causation something we observe?* Or is it merely postulated, perhaps in the interest of vindicating moral responsibility?[54] Unde-

---

52. My thanks to Timothy O'Connor for sending me some material from a manuscript currently in preparation. (The succeeding quotations are taken from this material.)

53. O'Connor has informed me that it is his view that there is indeed an *action* in such a case, but not an *agent-caused* action.

54. This is the view adopted by Randolph Clarke; see "Toward a Credible Agent-Causal Account," pp. 210–11.

niably, it does seem to many of us that we *do* observe agent causation—that, for instance, I observe *that I lift my arm*, and not merely *that I will for my arm to rise, and subsequently it does rise*. How is this disagreement to be resolved?

Probably what we say about agent causation at this point will parallel what we say about causation in general.[55] The principal motive for denying that we observe causation stems from a Humean perspective: a clever demon could bring it about that I seem to myself to perceive a causal sequence when I really don't, so what I *really observe* is not causation as such, but merely a sequence of "loose and separate" events. To be sure, on this Humean perspective I don't observe sticks and stones, either, but only "impressions" which I take to be derived from them. But once we've given up on the notion that observation ought to be infallible, these ideas tend to lose their grip. So perhaps it is reasonable to hold that we do observe rocks and trees, not merely impressions or sense-data. And perhaps we really do observe *that the rock broke the window*, and not merely *that the rock struck the window and thereafter the window shattered*. Why, then, should we not say we observe *that I raise my arm*, and not merely a sequence of noncausal facts?

The idea that we do observe agent causation does better than the contrary view in making sense of the compelling certainty we all experience with regard to the fact of our own agency. And yet, as we've admitted, this certainty is not proof against Humean skepticism, if the skepticism is strongly enough motivated. According to Richard Taylor,

> One can hardly be blamed here for simply denying our data outright, rather than embracing this theory [of agent causation] to which they do most certainly point. Our data—to the effect that men do sometimes deliberate before acting, and that, when they do, they presuppose among other things that it is up to them what they are going to do—rest upon nothing more than fairly common consent. These data might simply be illusions. It might in fact be that no man ever deliberates, but only imagines that he does, that from pure conceit he supposes himself to be the master of his behavior and the author of his acts. Spinoza suggested that if a stone, having been thrown into the air, were suddenly to become conscious, it

55. Interestingly, Clarke is following Michael Tooley, who denies in general that causal relations are observable.

would suppose itself to be the source of its own motion, being then conscious of what it was doing but not aware of the real cause of its behavior.[56]

I could hardly hope for a more dramatic confirmation of my suggestion that disbelief in agency ranks as a major form of skepticism, along with disbelief in the external world and in other minds. I hope to have persuaded some readers (but certainly not all!) that such skeptical heroics are unnecessary and unwarranted.[57] Why shouldn't we say, in the end, that we human beings are what we take ourselves to be and perceive ourselves to be—namely, free and rational agents?

56. Richard Taylor, *Metaphysics*, 2d ed. (Englewood Cliffs, N.J.: Prentice-Hall, 1974), p. 57.

57. For answers to some additional objections not explicitly considered here, as well as other valuable material, see O'Connor, "Agent Causation."

CHAPTER FIVE

Three Arguments for Substance Dualism

Rene Descartes once wrote:

First, since I know that all the things I conceive clearly and distinctly can be produced by God exactly as I conceive them, it is sufficient that I can clearly and distinctly conceive one thing apart from another to be certain that the one is distinct or different from the other. For they can be made to exist separately, at least by the omnipotence of God, and we are obliged to consider them different no matter what power produces this separation. . . . [S]ince on the one hand I have a clear and distinct idea of myself in so far as I am only a thinking and not an extended being, and since on the other hand I have a distinct idea of body in so far as it is only an extended being which does not think, it is certain that this "I"—that is to say my soul, by virtue of which I am what I am—is entirely and truly distinct from my body and that it can be or exist without it.[1]

Like Descartes, mind-body dualists have often wanted to produce metaphysical arguments for their theory. In this chapter we shall examine three such arguments in contemporary philosophy. The first two are adaptations of Descartes's argument given above; the third also has antecedents in Descartes, but appears more clearly in Leibniz and Kant.

1. Rene Descartes, *Meditations on First Philosophy*, trans. Laurence J. Lafleur (Indianapolis: Bobbs-Merrill, 1960), p. 74.

DESCARTES ACCORDING TO SWINBURNE

Richard Swinburne's modal argument for substance dualism, like that of Descartes,[2] focuses on the separability of mind from body.[3] The argument has attracted a good deal of critical attention, not all of it from materialists. Most of the criticisms allege that the argument is unsound, either because its premises are false or because it commits modal fallacies. I shall argue that even if the argument is sound it is epistemically circular and so provides no support for its conclusion.

Swinburne has stated his argument in several different ways, but the version he regards as most rigorous and definitive is found in Additional Note 2 to *The Evolution of the Soul*.[4] It goes as follows:

p = 'I am a conscious person, and I exist in 1984'
q = 'my body is destroyed at the end of 1984'
r = 'I have a soul in 1984'
s = 'I exist in 1985'
x ranges over all consistent propositions compatible with (p & q) and describing 1984 states of affairs. . . .

The Argument may now be set out as follows:
p Premise (1)

2. Swinburne states that he intended his argument to be an "improved version" of the one given by Descartes. See "Dualism Intact," *Faith and Philosophy* 13 (January 1996): 68.

3. Much of the material in this section comes from my "Swinburne's Modal Argument for Dualism: Epistemically Circular," *Faith and Philosophy* 15 (July 1998): 366–70; see also Swinburne's "The Modal Argument Is Not Circular," *Faith and Philosophy* 15 (July 1998): 371–72.

4. Page references in the text of this section are to Swinburne's *The Evolution of the Soul* (Oxford: Clarendon, 1986). It should be noted in this connection that William P. Alston and Thomas W. Smythe have charged Swinburne's argument with various modal fallacies ("Swinburne's Argument for Dualism," *Faith and Philosophy* 11 [January 1994]: 127–33); for Swinburne's reply see "Dualism Intact." As regards the argument as presented in chap. 8 of *The Evolution of the Soul*, their charges seem somewhat plausible. But Swinburne introduces the argument of Additional Note 2 explicitly "in case anyone suspects the argument . . . of committing some modal fallacy" (p. 314). Alston and Smythe's charge of a modal fallacy in this version of the argument, which rests on a particular interpretation of Swinburne's reason for affirming (2), is less than compelling.

$(x) \Diamond (p \ \& \ q \ \& \ x \ \& \ s)$ Premise (2)

$\sim \Diamond \ (p \ \& \ q \ \& \sim r \ \& \ s)$ Premise (3)

∴ ∼r is not within the range of x.

But since ∼r describes a state of affairs in 1984, it is not compatible with (p & q). But q can hardly make a difference to whether or not r. So p is incompatible with ∼r.

∴ r. (p. 314)

While this seems reasonably clear, our task of evaluation will be easier if we restate (2) spelling out the restrictions explicitly. 'x' will then be simply a propositional variable, and it becomes possible to formalize the entire argument. So we have:

1. p Premise

2'. $(x)\{[\Diamond (p \ \& \ q \ \& \ x) \ \& \ (x$ describes a state of affairs in 1984)]
      $\supset \Diamond (p \ \& \ q \ \& \ x \ \& \ s)\}$ Premise

3. $\sim \Diamond \ (p \ \& \ q \ \& \sim r \ \& \ s)$ Premise

4. ∼r describes a state of affairs in 1984   Premise

5. $\sim \Diamond \ (p \ \& \ q \ \& \sim r)$ From 2, 3, 4

6. $\sim \Diamond \ (p \ \& \ q \ \& \sim r) \supset \sim \Diamond \ (p \ \& \sim r)$ Premise

7. $\sim \Diamond \ (p \ \& \sim r)$ From 5, 6

8. r From 1, 7

This argument is unquestionably valid. There is, however, a difficulty about the interpretation of premise (3). On the face of it, this premise seems to entail that it is impossible for God to "re-create" a person who has been completely destroyed, as has been proposed by John Hick among others. But this seems difficult to reconcile with Swinburne's acceptance of the idea that it is possible for a substance to cease to exist and then come into existence a second time, and indeed that "it sometimes happens that a person (and so his soul) ceases to exist and then by an unexpected accident comes to exist again" (p. 179). Commenting on this, Swinburne states that "Premise 3 was not . . . meant to be understood in such a way that it ruled out re-creation; only re-creation of me without any part of me being re-created."[5] This is completely baffling. The re-creationist has no need to hold that a person is re-created without any part of him being re-created. On the contrary, the re-creationist will want to say that *the person's very own body* is re-created—either of the same elementary

5. Swinburne, "The Modal Argument Is Not Circular," p. 372.

particles or of others, as the case may be. If this possibility is not ruled out, then it is after all possible that p, and q, and $\sim r$, and s are all true together, and premise (3) is false.

Readers must make what they can of Swinburne's statements on this point. It does seem to me, however, that re-creationism is a conceptual absurdity,[6] so premise (3) will not be contested here. Premises (4) and (6) also seem correct, so the success of the argument will depend on the evaluation of (2').

In support of (2) (and (2')), Swinburne states, "Premise (2) relies on the intuition that whatever else might be the case in 1984, compatible with (p & q), my stream of consciousness could continue thereafter" (p. 314). Before we assess this, we need to look at an interesting objection suggested by Eleonore Stump and Norman Kretzmann. As a counterexample to (2) they proposed

I cease to exist at the last instant of 1984.[7]

Swinburne replied to this, "Any proposition which affirms that something existed throughout 1984 and then ceased to exist . . . clearly involves an entailment about a hard fact in 1985, viz., that there is no such thing then."[8] In saying this, Swinburne equates the idea of a proposition's "describing a state of affairs in 1984" with the idea of a proposition's asserting a hard fact about 1984, and he argues that the Stump and Kretzmann example can't be a hard fact about 1984 because it entails a hard fact about 1985. When in reply Stump and Kretzmann raised questions about the notion of "hard fact" employed by Swinburne, he replied in effect that, since it is his argument that is under review, he is entitled to his own definition of the notion of a hard fact. "So even if certain 'standard examples' of hard facts given by others don't count as hard facts on my definition, that is irrelevant to my argument which remains intact."[9]

Now, whatever the merits of this strategy may be, it has the consequence that it is out of the question to appeal to anyone's prephilosophical intuitions (as Swinburne wants to do) in support of (2) and (2'). For it

6. This will be argued explicitly in Chapter 8 below.

7. Eleonore Stump and Norman Kretzmann, "An Objection to Swinburne's Argument for Dualism," *Faith and Philosophy* 13 (July 1996): 406.

8. See ibid., p. 407. Note that this once again contradicts Swinburne's acceptance of the idea that a substance can cease to exist and then again begin to exist.

9. "Reply to Stump and Kretzmann," *Faith and Philosophy* 13 (July 1996): p. 413.

is wholly implausible to suppose that a philosophical layperson, or even a philosopher who has not specifically considered the matter, has in mind Swinburne's idiosyncratic notion of a "hard fact," and uses this in forming an intuition about the truth of (2). And we can't rely here on some general, pre-analytic idea of what it is for a proposition to describe a state of affairs about a particular time. As the Stump and Kretzmann counterexample shows, the precise explication of "hard fact" is crucial.

But if prephilosophical intuitions aren't in play here, what about the intuitions of professional philosophers? I strongly suspect that the vast majority of us, at any rate, are no better off than the layperson with respect to grasping intuitively the truth or falsity of (2). For one thing, it is not clear to me that Swinburne has specified his notion of "hard fact" sufficiently to enable us to know what its necessary and sufficient conditions may be, let alone grasp it intuitively so as to use it in evaluating (2).[10] And (2) (better, (2')) is in any case a moderately complex proposition; the sort of proposition, in fact, concerning which it behooves one to be quite modest about his ability to assess its truth by logical intuition.

I suggest, then, that if we are to evaluate (2') we can do no better than take the course marked out for us by Stump and Kretzmann, and proceed by first proposing and then evaluating possible counterexamples. And for the time being, it may be best to avoid counterexamples whose "hard fact" status is in question. If a clear counterexample should emerge (2') is doomed, whereas if (2') proves robust in the face of putative counterexamples it is at least worthy of serious consideration.

But now, what about $\sim r$ itself as a counterexample to (2')? What has to be evaluated is the following instance of (2'):

$$(2'^*)[\Diamond\,(p\ \&\ q\ \&\ \sim r)\ \&\ (\sim r\ \text{describes a state of affairs in 1984})] \supset \Diamond\,(p\ \&\ q\ \&\ \sim r\ \&\ s)$$

We have already agreed to premise (3), which states that the consequent of (2'*) is false, and to (4), which states that the second conjunct in the antecedent is true.[11] So the counterexample will succeed if and only if $\Diamond\,(p\ \&\ q\ \&\ \sim r)$ is true. Now $\Diamond\,(p\ \&\ q\ \&\ \sim r)$ can be true only if $\Diamond\,(p\ \&\ \sim r)$ is

---

10. For some of the perplexities, and contestable assumptions, involved in Swinburne's employment of this notion, see the articles by Stump and Kretzmann and Swinburne cited above.

11.Of course, if either of these premises were rejected, the argument would immediately be rejected as unsound.

also true. And, given (6), the converse also is true; if ◊ (p & ~r) is true so is ◊ (p & q & ~r). So (2') survives the alleged counterexample only if ◊ (p & ~r) is false and (7) ~◊ (p & ~r) is true instead. Now, opinions about (7) will certainly vary. Swinburne thinks it is true; materialists, and a good many dualists other than Swinburne, will think it false. The important point, however, is this: *(7) is itself a principal conclusion of Swinburne's argument for dualism.* To be sure, the formal conclusion of the argument is not (7) but rather (8) r. But the move from (1) and (7) to (8) is trivial; Swinburne could just as well have left (7) as the final conclusion (thus, incidentally, rendering premise (1) superfluous). But this means that Swinburne's argument for dualism is *epistemically circular* in the sense explicated by Victor Reppert,[12] namely, that "no reasonably well-informed person would accept the premise who does not already accept the conclusion."[13] Indeed, any reasonable person must acknowledge that the premise in this case is *less* well-supported than the conclusion, since even if (7) is true there are additional serious questions that can arise about the truth of (2')—for instance, questions about Swinburne's definition of hard facts, as well as about other possible counterexamples to (2'). It seems, then, that Swinburne's argument can provide no rational support whatever for the acceptance of dualism.[14]

12. See Victor Reppert, "Eliminative Materialism, Cognitive Suicide, and Begging the Question," *Metaphilosophy* 23 (1992): 389 (and see the discussions in Chapter 1).

13. Swinburne objects to this that "it does not seem very plausible to suppose that the argument is epistemically circular in the stated sense—since someone might accept the premises [*sic*] without ever understanding the conclusion," which "concerns an item not mentioned in this premise—'a soul' " ("The Modal Argument Is Not Circular"). It would seem doubtful that anyone can accept *all* the premises of Swinburne's argument without understanding the concept of a soul! More fundamentally, however, my difference with Swinburne has to do with what makes someone a "reasonably qualified person" in this context. The epistemically circular argument discussed in Chapter 1 had to do with someone who wanted to prove God's existence on the basis that the Bible, which contains no falsehood, says that God exists. Conceivably, someone might accept that everything the Bible says is true because Uncle Harry, who is a really nice man, says that about the Bible. Such a person would not, according to Reppert, count as a "reasonably qualified person." Nor would I recognize as reasonably qualified someone who might accept Swinburne's premise 2 without even suspecting that he was trying to prove something about an immaterial part of the person.

14. This is by no means merely the point that, as Swinburne acknowledges, his "argument will not convince anyone who claims to be more certain that the conclusion is false than that the premises are true" ("Dualism Intact," p. 71). As Swinburne

It might seem that the conclusion about epistemic circularity could be avoided if someone were to have a clear and specific rational intuition about the truth of (2') which did not depend on the belief that p and $\sim r$ are incompatible. But how is this supposed to go? Regardless of the intuitions that may be involved, anyone possessing modest logical acumen is bound to recognize the need to test (2') against possible counterexamples.[15] And given the content of the argument, the possibility that $\sim r$ might constitute a counterexample lies ready to hand. But once this possibility has been raised, the game is up: either the counterexample succeeds, and the argument is unsound, or it fails, and the argument is epistemically circular.[16] So Swinburne's argument is *at best* epistemically circular, and it contributes nothing to the rational acceptability of dualism for anyone, *including Swinburne himself.*

DESCARTES ACCORDING TO TALIAFERRO

For another contemporary example, we turn to Charles Taliaferro's *Consciousness and the Mind of God*.[17] Taliaferro's argument is on the whole closer to Descartes's own version than is Swinburne's; as we

---

rightly states, this is true of all arguments and constitutes no particular fault of his argument for dualism. But it is not true of all arguments—or of any good ones—that they are epistemically circular, in that the premises cannot reasonably be accepted by anyone not already convinced that the conclusion is true.

15. Swinburne, to be sure, disagrees with this. He writes, "The grounds for accepting the premises are the coherence of various thought experiments described in two pages of my text [viz., pp. 151–52]; including ones easily graspable by seven-year old religious believers or readers of fairy stories." Swinburne seems to me to have been unusually fortunate in his experience with seven-year-olds. I've known some fairly bright children of that age, but I've never encountered one I thought capable of grasping Swinburne's definition of a hard fact, or of comprehending a premise such as (2) which presupposes an understanding of this definition.

16. According to Stump and Kretzmann, Swinburne considers it question-begging for an opponent to invoke $\sim r$ as a counterexample to (2) (Stump and Kretzmann, p. 411 n. 9). Whatever the merits of this contention, it has no force against my procedure in this discussion. I don't assert that $\sim r$ is a counterexample; instead, I show that whether it is or not, the argument is ineffective.

17. Cambridge: Cambridge University Press, 1994. (Page references in this section are to this book.)

shall see, it also permits a clear vision of the difficulties confronting this sort of argument.[18] Taliaferro states the argument as follows:

Let "A" refer to me and "B" to my body.
1. A is B (the hypothesis of the identity materialists).
2. A cannot exist without B and B cannot exist without A.
3. But A can exist without B and B can exist without A.
4. Therefore premise "1" is false; it is not the case that A is B. (p. 175)

Here (2) is derived from (1) by way of a principle with impeccable credentials, namely Leibniz's Law, or the indiscernibility of identicals. Since (3) contradicts (2), it is perfectly correct to infer from (2) and (3) that (1) is false, and the person in question, namely A, is not identical with his body, B. So the only point at which questions could be raised concerns (3), in particular the assertion that it is possible for the person to exist without his body. It is important here to see that Taliaferro's (3) does not content itself with the *de dicto* modal claim that the proposition "I exist without a body" is not self-contradictory. Rather, (3) identifies both the person and the body in question by direct reference, and attributes to the person the *de re* modal property of being such that it is possible for it to exist without its body. If the claim that the person actually possesses this property can be supported, the argument will go through.

But what reason is there to think that (3) is true? The answer is given by thought-experiments involving body-transfer (*without* brain transplants), out-of-body experiences, and personal survival after death. Insofar as these thought-experiments can be imagined in convincing detail (and a number of examples are given to show that this can be done), they give good reason to suppose that existence apart from one's body is at least metaphysically possible—that it occurs in some possible world, even if not in the actual world. This accords with Taliaferro's "principle of strong conceivability," which he formulates as follows: "*A subject, S, is justified in believing that a state of affairs is possible if the state of affairs seems possible to S and S is not intellectually negligent*" (p. 157).[19]

I believe the principle of strong conceivability is misapplied if it is used

18. For another, more extensive, development of this type of argument, see W. D. Hart, *The Engines of the Soul* (Cambridge: Cambridge University Press, 1988), especially chaps. 1–4.

19. Taliaferro explains further: "Strong conceivability involves a variety of factors, including forming a clear and distinct picture of the state of affairs or providing a de-

to support a premise such as (3).[20] In order to know that this premise is true we require insight into the essential properties of concrete individuals, and there are many cases in which no amount of intellectual diligence will give us such insight. Suppose the desk I used to have in my office was made of oak. If that is so, then there is no possible world in which *that desk* exists made of maple. But since I don't *know* that it was made of oak, no amount of intellectual diligence, conceptual analysis, and the like can reveal to me the impossibility of its being made of maple.[21] Then there is the famous case of water, which in fact is essentially H_2O but which we can perfectly well conceive of as being XYZ instead. And it is often pointed out that we can construct thought experiments in which persons are not separable from their bodies as well as thought experiments in which they are separable.

The root difficulty here seems to be this: a thought-experiment may show me that *I have no reason to believe* that a given state of affairs is impossible—that is, it shows me that the state of affairs in question is *epistemically possible*. But in order to be sure, on this basis, that the state of affairs is *metaphysically possible* I should have to know that I correctly understood the essential natures of the individuals in question. But in the case of (3) the essential nature of human beings is just what I am trying to ascertain. So inferring the truth of (3) from thought-experiments involves either a confusion of epistemic possibility with metaphysical possibility or an illicit inference of the latter from the former.

Now, Taliaferro is by no means unaware of possible objections to his thought-experiments; indeed, he spends a good many pages answering such objections. It is not possible here to respond to everything he says, so I will select three of his points for further comment. The first response can be formulated as follows: *Admittedly, the modal judgments involved in the thought-experiments are neither infallible nor incorrigible; nevertheless, they do possess considerable weight as support for dualism.* In principle, I agree with the underlying idea here: I too am an epistemological fallibilist, and

----

tailed description of it and comparing this closely with one's picture and description of the world" (p. 157).

20. I believe, on the other hand, that thought-experiments do help to create a strong case for *property* dualism—what Taliaferro calls "dual-aspect theory."

21. W. D. Hart discusses this type of objection; so far as I can tell, his view is that there is indeed a possible world in which the very same desk exists made of maple. See his *Engines of the Soul*, pp. 34–40.

I hold that fallible human intellectual processes (the only kind we possess) can provide us with warranted beliefs and even with knowledge. But I have my doubts whether the thought-experiments in support of (3) can provide any warrant whatever in support of a modal argument for substance dualism. We have to keep in mind that the sort of modal reasoning in question here is a type of reasoning in which we seem to be especially susceptible to subtle fallacies of various kinds. Some of these fallacies have already been mentioned; probably the most insidious is the conflation of epistemic possibility with metaphysical possibility. Insofar as one's inclination to accept (3) may be influenced by one or another of these fallacious forms of reasoning, that inclination provides no support whatever for (3) or for the modal argument. On the other hand, one might have a genuinely warranted belief in the truth of (3) derived from an entirely rational prior acceptance of, or inclination toward, mind-body dualism. But while one's belief in (3) in such a case might be warranted, a modal argument for dualism having this as its basis would be epistemically circular and thus ineffective. In order to provide a basis for the modal argument, one's acceptance of (3) must be based on a genuine modal intuition which is derived neither from the influence of fallacious modal reasoning nor from one's prior inclination to accept mind-body dualism.[22] There is no way to prove that someone might not have valid intuitions of that sort, but I for one remain deeply skeptical.

A second point made by Taliaferro in defense of his argument attempts to undermine the alleged parity between dualist and antidualist thought-experiments. In brief, his claim is that *whereas dualist thought-experiments leave no doubt that the person is not identical with her body, antidualist thought experiments do not clearly show the identity of the person with her body.* As a lead-in to this argument, consider once again the famous Water–H_2O–XYZ case. Taliaferro suggests that we cannot, as alleged, easily construct thought-experiments in which water exists but is not H_2O. "What is involved with conceiving of water without conceiving of H_2O? A hydrogen or oxygen atom hardly admits of easy visual detection, and if asked to imagine either, many people would probably draw a blank. Lacking such a conception, it is difficult for us to follow instructions asking us

---

22. Admittedly, one might have an inclination to accept person-body separability without having any explicit philosophy of mind at all. And then one might be influenced by the modal argument toward an explicit acceptance of mind-body dualism. But this is not the use of the modal argument that is in question here.

to imagine two cases, one in which water is composed of H_2O and another in which it is not" (p. 190).[23]

Similar difficulties, Taliaferro alleges, attend the attempt to construct a thought-experiment exhibiting mind-body identity:

> To imagine that you are your body will have to involve more than imagining that you are volitionally, sensorily, and proprioceptively embodied. Distinguishing identity from integrative cases will also have to involve more than imagining that if your body perishes you do, for it is perfectly compatible with integrative dualism [Taliaferro's own view] that persons do not survive the demise of their body. I believe that to picture the person-body identity successfully, one must imagine that something could not be the case, namely that it is impossible that the person can become disembodied or switch bodies. . . . This involves a negative existential proposition of a high order. (p. 213)

Taliaferro clearly thinks claims to "imagine" such a negative existential deserve to be met with skepticism.

Contrary to Taliaferro, I think it may not be too difficult in either case to craft thought-experiments which restore parity. The case of water is relatively simple: Imagine a liquid which in all everyday contexts behaves exactly like water (rain and snow, drinking, swimming, and watering plants), but which gives different results when subjected to chemical tests convincing us that water is, in fact, H_2O. Perhaps it decomposes under electrolysis into two different gases, neither of which is either hydrogen or oxygen.

My favorite materialist thought-experiment comes from Peter van Inwagen's book *Metaphysics*. Imagine you have a "duplicating machine" which, when an object is placed in one of its two chambers and a button is pressed, produces a physically indistinguishable duplicate in the other chamber. After duplicating the Hope diamond, the British crown jewels, the *Mona Lisa*, and a few other such memorabilia, you try the experiment of duplicating a living human being. Clearly, the predicted result for materialism differs from that for certain dualist theories. According to

---

23. In all fairness it should be stated that this is Taliaferro's second response to this example; his preferred response is that "if you believe you can conceive of water and not H_2O and have carefully reviewed the counter-evidence available to you, then you do possess *prima facie* warrant in believing there to be a distinction between the two" (p. 169)—a warrant, however, which is overridden by scientific findings.

Descartes, a living human body would appear in the second chamber, but "this body would immediately crumple to the floor. It would lie there breathing and perhaps drooling, and, if you force-fed it, it would digest the food and in time produce excreta. But it would not *do* anything much."[24] A Muslim student of van Inwagen's, on the other hand, thought that the experiment would produce a corpse, since the soul, which is the principle of life, would not be reproduced. Van Inwagen's own view (and the one he expects most of his readers to agree with) is that the result would be a living person indistinguishable from the original. Van Inwagen, in fact, terms this the "single argument for physicalism that I find the most powerful and persuasive."[25]

Taliaferro, however, has claimed in discussion that this thought-experiment, even with the result as predicted by van Inwagen, would not be fatal to all versions of dualism.[26] And this seems correct; it would not, for example, disprove the emergent dualism advocated in this book. For that matter, the dualist could always hypothesize that God had created a soul for the duplicate body. The difficulty seems to be this: When one is dealing with an entity that, by hypothesis, is completely imperceptible, there is no decisive way to *imagine* either the presence of that entity or its absence. But one can *conceive* of the existence or nonexistence, the presence or the absence, of many things one is unable either to perceive or to imagine. And *conceiving* of a state of affairs is just as good (or as bad) a reason for thinking it possible as is *imagining* it; to think otherwise is to indulge in a sort of "modal verificationism." If the existence of the soul is held to be possible because conceivable (even though it cannot be imagined), then equally the materialist thought-experiments are possible because conceivable, even though their *imaginative content* is not distinct from that of all possible varieties of dualism.

Even were he to grant van Inwagen's thought-experiment, Taliaferro might still claim an advantage for the dualist thought-experiments simply because there are so many actual reports of out-of-body experiences. Thus, he writes, "The difference between the water and person-body cases can be brought out by taking note of the enormous amount of parapsychical literature purporting to document cases of disembodiment and

24. Peter van Inwagen, *Metaphysics* (San Francisco: Westview, 1993), p. 181.
25. Ibid., p. 180.
26. Taliaferro argues at length for the nonparity of dualist and materialist thought-experiments in "Possibilities in Philosophy of Mind," *Philosophy and Phenomenological Research* 57 (March 1997): 127–37.

body switching. We do not have a similar class of reported incidents in which people claim to see and drink water, discovering that it is not H_2O but XYZ!" (p. 191).

What are we to make of the prevalence of belief in, and reports of, out-of-body experiences and the like? Certainly they suggest that a fairly large segment of the human population entertains views that are more consonant with dualism than with materialism. And from a certain standpoint, this may constitute significant support for dualism. On the other hand, it can't be expected to move a sophisticated materialist, who may be quite willing to acknowledge that untutored human thinking tends to bend in the direction of dualism. It would be a different matter, of course, if one were to argue that the reports in question are actually true, and that they really do recount the experiences of persons while separated from their bodies. But Taliaferro doesn't seem inclined to make that move[27]—and were he to do so, we should have moved beyond modal arguments for dualism into considerations of a quite different nature.

THE UNITY-OF-CONSCIOUSNESS ARGUMENT

Ever since Descartes there have been philosophers who have claimed that the unity of conscious experience argues strongly against the possibility that the mind or self is a material thing.[28] This argument surfaced from time to time during the seventeenth, eighteenth, and nineteenth centuries, but recently it has been neglected. My contention is that this ne-

27. He says, "I would go further and claim that parapsychology gives us reason to believe in the actual survival of some people after biological death for a brief period of time, but for now I only appeal to OBE reports as backing up and filling out what might otherwise be a merely academic thought experiment" ("Animals, Brains, and Spirits," *Faith and Philosophy* 12 [October 1995]: 372).

28. "For in reality, when I consider the mind—that is, when I consider myself in so far as I am only a thinking being—I cannot distinguish any parts, but I recognize and conceive very clearly that I am a thing which is absolutely unitary and entire. . . . But just the contrary is the case with corporeal or extended objects, for I cannot imagine any, however small they might be, which my mind does not very easily divide into several parts, and I consequently recognize these objects to be divisible. This alone would suffice to show me that the mind or soul of man is altogether different from the body, if I did not already know it sufficiently well for other reasons" (*Meditation* 6, pp. 81–82).

glect is a mistake, and that the argument places a serious and perhaps insuperable obstacle in the way of materialist theories of the mind.

Kant's "Paralogism"

The argument against materialism from the unity of consciousness is clearly present in Leibniz, as Margaret Wilson has shown.[29] Indeed, the classic form given the argument in the Second Paralogism may well be viewed as Kant's reflective development of the argument taken from Leibniz. Kant's version goes as follows:

> Every *composite* substance is an aggregate of several substances, and the action of a composite, or whatever inheres in it as thus composite, is an aggregate of several actions or accidents, distributed among the plurality of the substances. Now an effect which arises from the concurrence of many acting substances is indeed possible, namely, when this effect is external only (as, for instance, the motion of a body is the combined motion of all its parts). But with thoughts, as internal accidents belonging to a thinking being, it is different. For suppose it be the composite that thinks: then every part of it would be a part of the thought,[30] and only all of them taken together would be the whole thought. But this cannot consistently be maintained. For representations (for instance, the single words of a verse), distributed among different beings, never make up a whole thought (a verse), and it is therefore impossible that a thought should inhere in what is essentially composite. It is therefore possible only in a *single* substance, which, not being an aggregate of many, is absolutely simple.[31]

In dubbing this a "paralogism," Kant seemingly invites us to regard it as wholly unsound—yet his own attitude toward the argument was far more

29. Margaret D. Wilson, "Leibniz and Materialism," *Canadian Journal of Philosophy* 3 (June 1974): 495–513.

30. It is difficult to make sense of the notion that each part of the composite would *be* a part *of the thought*. Roderick Chisholm translates the passage in question as follows: "Suppose a compound thing were to think. Then every part of that compound would have a part of that thought. The thought that the compound would then have would be composed of the thoughts of the parts of that compound" ("On the Simplicity of the Soul," *Philosophical Perspectives* 5 [1991]: 175).

31. Immanuel Kant, *Critique of Pure Reason*, trans. N. Kemp Smith (New York: St. Martin's, 1965), p. 335 (A 352).

complex than that.[32] We must remember, to be sure, that the chief objective of Kant's "transcendental psychologist" is to establish a speculative proof of immortality. And it clearly is Kant's view that neither this argument nor any other can succeed in doing this. But in certain other respects, his assessment of the unity-of-consciousness argument is far from negative. On the contrary, in the second edition of the *Critique* he wrote, "From this [argument] follows the impossibility of any explanation in *materialist* terms of the constitution of the self as a merely thinking subject."[33] And the same claim is repeated in *On the Progress of Metaphysics*: "therefore materialism can never be used as a principle for explaining the nature of the soul."[34] Henry E. Allison summarizes Kant's attitude toward the argument as follows: "Thus, Kant's position seems to be that the unity of consciousness, which the rational psychologist (presumably Leibniz) uses erroneously to establish the positive metaphysical doctrine of the simplicity and hence incorruptibility of the soul, can be used legitimately to establish the weaker thesis of the impossibility of a materialist explanation of the conceptual activities of the mind."[35]

It has to be said that Kant's partially favorable attitude toward the argument has not been shared by his recent commentators and critics.[36] But rather than pursue this debate in the Kantian context, I think it will be

---

32. See Karl Ameriks, *Kant's Theory of Mind* (Oxford: Clarendon, 1982), chap. 2.

33. Kant, *Critique of Pure Reason*, p. 376 (B 420).

34. *Kants gesammelte Schriften*, IV, p. 308; quoted by Henry E. Allison, "Kant's Refutation of Materialism," *Monist* 79 (April 1989): 195.

35. Allison, "Kant's Refutation of Materialism," p. 195. Even this may be overly simple, however. For a detailed discussion of Kant's view of the argument, see Ameriks, *Kant's Theory of Mind*.

36. See Wilson, "Leibniz and Materialism," pp. 509–12; Ameriks, *Kant's Theory of Mind*, pp. 55–64; Allison, "Kant's Refutation of Materialism," p. 18. Roderick Chisholm, who is not unsympathetic to the argument, nevertheless rejects it because, in his opinion, we lack a sense for the expression "a part of a thought," and thus "have no reason to accept the statement" in Kant's argument which employs this expression (Chisholm, "On the Simplicity of the Soul," p. 175). But this is far from conclusive. If we think of the "thoughts" in question as propositions, then it seems rather natural (though it is not uncontroversial) to think of the concepts involved in the proposition as "parts" of the thought. But, more important, the argument does not actually need to make use of the notion of a "part of a thought." What *is* needed is merely that a person's being aware of a complex cannot consist of the actions of parts of the person each of which is *not* aware of that complex.

most helpful for us to move directly to a contemporary version of the argument.[37]

Sellars and the Principle of Reducibility

First, we begin with some stipulations about the materialism against which the argument is directed. This materialism will be assumed to view the brain as a computerlike network of electrochemical interconnections, processing sensory (and other) data much in the way a computer processes information. Each item of information held within the brain is modeled in the physical state of some portion of the brain. There is not conscious awareness of every item of information the brain contains, but awareness when it occurs consists of the brain, or some part thereof, being in the appropriate physical state. An important question here, still in dispute among brain scientists, is whether items of information are stored in discrete subunits of the brain, or whether information is stored "holographically"[38] so that any given item is spread over a large region of brain tissue. In what follows, this question will be left open.

As an example of the unity of consciousness, I cite my awareness of my present visual field.[39] This field includes the impressions from a set of shelves in the living room of my apartment, with books on the lower shelves and a number of plants, mostly miniature cacti, above them, while behind the shelves there is the horizontal pattern of the mini-blinds. All this I observe without scanning or refocusing my eyes: momentarily, as it were. This visual field is not unified in any interesting aesthetic sense, but

37. Some of the material in the next few pages has been adapted from my article, "Emergentism," *Religious Studies* 18 (1982): 480–84.

38. On a holographic film information from any point in the field of view is distributed all over the film, rather than being concentrated on a single point as in an ordinary photographic negative. If a part of a holographic film is destroyed, no part of the field of view is lost; instead, the quality of the entire image is degraded to some extent.

39. The term "visual field" as here employed is perhaps not quite everyday usage, so some clarification may be in order. We certainly are familiar with the notion that each of us has at any given time a "field of vision" such that, for example, unobstructed objects in front of one are in one's field of vision but may leave the field of vision by moving behind one's back. The "visual field" is simply one's field of vision insofar as one is aware of it at a given time, assuming one is not asleep, one's eyes are open and functioning, and so on.

it is a fact that I experience it as a unity, all at once and not as a succession of discrete experiences.[40]

Before proceeding further, it will be helpful to see what our datum looks like as viewed from the perspectives of two contemporary schools of materialism. According to functionalism, my being aware of my conscious field is my being in a state with certain causal and functional properties. And according to eliminativism, what I describe as my being aware of my visual field is a matter of certain parts of my brain being in certain physical states. Now, it may well be that my being aware of my visual field has exactly the causal and functional properties attributed to it by functionalism. And it may also be true that when I take myself to be aware of my visual field my brain is in such-and-such a physical state. But what I *mean* when I say that I am aware of my visual field is *not* that I am in a state with certain causal and functional properties, or that my brain is in such-and-such a physical state. The meaning of the assertion that I am aware of my visual field is, I think, neither obscure nor difficult to grasp.[41] Furthermore, I believe that, on many occasions, what is thus asserted by me is *true*. The reader, then, is invited to take this datum at face value and to investigate it in its own right, without first translating it into the argot either of functionalism or of eliminativism.[42]

---

40. I wish to emphasize that the "unity" in view here is the mere fact of being experienced simultaneously by a single subject; it does not involve any of the more sophisticated (and possibly debatable) ways in which some philosophers (e.g., Kant) have alleged that experience is and must be unified. One might also consider as a datum the "unified" experience of simultaneously feeling a tickle, hearing a twig snap, and smelling gasoline.

41. But might not the mental property I thus take myself to have turn out to be identical with some (perhaps rather complex) physical property? I am not at this point concerned to decide this question; I merely assert that I do in fact sometimes have the property of being aware of my visual field.

42. Does this argument, then, presuppose the rejection of eliminativism and functionalism as theories of the mind? As noted, it is consistent with what has been said so far that my being aware of my visual field has exactly the causal and functional properties attributed to it by functionalism, and that when I take myself to be aware of my visual field my brain is in exactly the physical state postulated by eliminativism. According to eliminativism, however, the statement "I am aware of my visual field," like other statements in "folk psychology," is *not true*. Since I believe (and have argued in Chapter 1) that this and similar statements are often true, I do reject eliminativism. It is not so clear that functionalism must reject commonplace assertions about the mental as false. What is clear is that on the view taken here the causal-functional story

Now, my procedure is to take a specific conscious state—the state I am in when I am aware of my visual field, as described above—and ask what physical entity it is that is in that state. The question is surprisingly difficult for the materialist to answer. Initially, to be sure, the materialist may reply that the physical object which is aware of my visual field is *I myself*— that is, my body. But not all of my body is equally relevant to the state in question; my brain, we generally believe, is essential to it in a way that other parts of my body are not. (The brain could be in a similar state even if it were in a vat.) So is it my brain, rather than I, that is aware of the visual field? This raises a question of terminology. When a person performs some action in virtue of some part of that person's body doing something, we sometimes say that the *person* has done something *by means of* the bodily part, and sometimes that the *part itself* has done it. (I point *with my finger*, but also *my finger points*.) It is not clear, in the case of mental states such as thinking and awareness, that one of these ways of speaking is more correct than the other: Do I "use my brain" to figure things out, or does my brain figure things out?[43] In what follows, the latter form of speech will be employed, but all the points made will apply in either case.

Let us say, then, that it is my brain that is aware of my visual field, and I am aware of it in virtue of my brain's being aware of it. But not all of my brain need be involved—some of it may be performing other functions which contribute indirectly or not at all to my awareness of my visual field. So let us narrow the focus even further. Let V be the smallest part of my brain which contains the modeling of all of the information from my visual field. The existence and functioning of V may or may not be sufficient for my having the awareness of the field. But it clearly is necessary for the awareness, since if part of V is not functioning some of the information-content of the field will be lost.

Should we say, then, that it is V which is aware of the visual field? Perhaps so, but we need to consider the composition of V. In the light of the computer analogy sketched above, we can say that V is a whole composed of physical parts. Many of these parts model information from various parts of the visual field. But no proper part of V models all of this infor-

---

told by functionalism cannot be the *whole truth* about the mind. As Searle says, mental states have an "irreducibly subjective ontology."

43. A dualist, of course, would have a metaphysical reason for preferring the former locution. But in the present context we are assuming the perspective of the materialist, who would not have a reason of this kind.

mation, so it is not possible for any of these parts to be aware of the entire visual field. But if V is a whole composed of parts each of which is *not* aware of the visual field, how can V itself be aware of it? If we assume that each item of information is modeled in a discrete subunit of the brain, we might suppose that each subunit is aware of the information it contains, and that, in virtue of this, V is aware of the entire field. (On the holographic model, on the other hand, *all* of V is needed to carry the full information for *any part* of the visual field.) But even if the subunits are assumed to be aware of the items of information they contain, this does *not* enable us to explain the awareness of the entire field; this would be like saying that each student in a class knows the answer to one question on an examination, and that in virtue of this the entire class knows the material perfectly! It is true that the members of the class are able, working together, to *reproduce* all of the information, but there may in fact be no one at all who *knows* or *is aware of* all of it. The point is simply that the kind of awareness we are discussing is essentially unitary, and it makes no sense to suggest that it may be "parceled out" to entities each of which does *not* have the awareness. *A person's being aware of a complex fact cannot consist of parts of the person being aware of parts of the fact.* A conjunction of partial awarenesses does not add up to a total awareness.[44]

But may we not suppose that, in addition to the various subunits which model items of information from the field, V also contains a scanning mechanism, S, which collects and integrates the items of information and thus makes possible the unified awareness? There seems in fact to be a general recognition that no such "synthesis organ" for consciousness actually exists.[45] But even if we were to postulate such a mechanism, it

44. But suppose the brain includes a number of parts, each of which contains *all* the information from the visual field? In this case we have, not a *single* V, but *many* V_i, *many* brain-units, each of which might conceivably (according to the argument thus far) be aware of the entire visual field. Now, take any one of these—say, V_n—and consider its composition. In the light of the computer analogy sketched above, we can say that V_n is a whole composed of physical parts. Many of these parts model information from various parts of the visual field. But no proper part of V_n models all of this information, so it is not possible for any of these parts to be aware of the entire visual field. But if V_n is a whole composed of parts each of which is *not* aware of the visual field, how can V_n itself be aware of it? And with this question, we return to the main argument; the postulation of multiple V_i doesn't change the situation at all.

45. Consider the following from Owen Flanagan: "One mistake to stay away from is positing a center of consciousness, a specific faculty devoted to consciousness,

would not solve the problem pointed out here. Note that for this mechanism to do what is required of it, *S* must somehow represent within itself *simultaneously* all of the information of the visual field. Now if *S* does contain all of this information (even if only momentarily), we must still reflect that *S* is itself a whole consisting of parts—and thus the argument begins all over again, with the same result as before.

I think what one naturally wants to say at this point is something like the following: It may be, indeed must be, that the information from my visual field is distributed among a number of distinct brain-units. But this fact obviously does not make a unified consciousness impossible—it would, after all, be absurd to require that all of the information should somehow be concentrated into a mathematical point! Rather, what happens is that the various items of information are as it were drawn together and co-presented in a unified consciousness, and this process is not frustrated by the spatial "spread" of information within the brain.

It seems to me that something like this must be true. Note, however, that "consciousness," as used in the preceding paragraph, does not refer to a brain or to any part of a brain. So I repeat my question: what *physical entity* is it that is aware of my visual field? If materialism cannot answer this question, it is in serious difficulty.[46]

For many materialists, the most appealing solution may lie in rejecting the demand, which is implicit in the discussion to this point,[47] that the properties attributed to the brain be accounted for in terms of properties of, and relations between, its parts. For an especially clear formulation of this demand, we turn to Wilfrid Sellars and his "principle of reducibility," which he states as follows:

---

that receives some but not all messages causally relevant in human activity. Some patterns of neural activity result in phenomenological experience; other patterns do not. The story bottoms out there" (*Consciousness Reconsidered* [Cambridge: MIT Press, 1992], p. 58). And see Daniel Dennett's remarks about the "Cartesian Theater" in chapter 6 of *Consciousness Explained* (Boston: Little, Brown, 1991).

46. If at this point we revert to saying that the *person as a whole* has this awareness, an awareness that is not possessed by any part of the person, we confront once again the truth that a person's being aware of a complex fact cannot consist in the actions of parts of the person, each of which does *not* possess this awareness.

47. And in Kant's argument: "The actions of a composite, or whatever inheres in it as thus composite, is an aggregate of several actions or accidents, distributed among the plurality of the substances."

If an object is *in a strict sense* a system of objects, then every property of the object must consist in the fact that its constituents have such and such qualities and stand in such and such relations, or roughly, every property of a system of objects consists of properties of, and relations between, its constituents.[48]

Commenting on this principle, Sellars states: "Telling us, as it does, that if an object *is* (as contrasted with *is correlated with*) a whole of parts, its having *P* consists in its parts having properties and standing in relations, it also tells us that if an object has a property which violates the principle, then *in that context* it is *correlated with* rather than *consists of* the 'parts.' "[49]

Three points need to be noted with regard to Sellars's principle of reducibility. First, Sellars's name for his principle may suggest the notion of reduction as employed in the philosophy of science, but it would be a mistake to take it that way. This is a *metaphysical* principle, belonging, as Sellars says, "to logic or general ontology rather than to the philosophy of science."[50] Second, since the intention of the principle is metaphysical rather than epistemological, what is required is that the properties of the system shall consist in properties *actually possessed* by the parts, not necessarily in those *we know* (or believe on independent evidence) to be possessed by them. (Thus, in some cases the requirements of the principle might be satisfied by postulating new, previously unknown, properties for the parts.) Third, the principle deals with the relations between the properties of a system and properties of its parts *at a given time*; it is, one might say, synchronic rather than diachronic.

While I have little desire to plunge into a maze of Sellarsian exegesis, his expression "consists of" (or "in") requires some comment. I take it that Sellars is describing a situation in which a whole is *nothing but* a system of objects—where there is no such thing as a whole "over and above" the sum of the parts—in distinction from a situation in which there is such a whole that has properties of its own that cannot be accounted for in terms of the parts and their properties. The phrase can be helpfully illustrated by

48. Wilfrid Sellars, "Philosophy and the Scientific Image of Man," in *Science, Perception and Reality* (London: Routledge & Kegan Paul, 1963), p. 27.

49. Wilfrid Sellars, "Science, Sense Impressions, and Sensa," *Review of Metaphysics* 24 (1971): 411. Concerning the justification for this principle, Sellars wrote, "I have as yet published no . . . defense, but have simply stated that I find the principle in accordance with my logical intuitions" (p. 393).

50. Ibid., p. 411.

Sellars's own examples. The redness of a brick wall *consists in* the redness of its constituent bricks—the redness of the wall amounts to nothing over and above the fact that the visible surface of the wall is made up of the surfaces of the bricks (disregarding mortar lines), and those surfaces are red. And for a certain object to be a ladder "is for its parts to be of such and such shapes and sizes and to be related to one another in certain ways."[51] In contrast, the pinkness of a pink ice cube does *not* consist in its constituent elementary particles having certain properties and standing in certain relations, for none of the particles have properties of which "being pink" could consist. (This helps to motivate Sellars's well-known contrast between the "manifest image" and the "scientific image" of man-in-the-world, as well as the doctrine that the scientific image stands to the manifest image as reality to appearance.)

But perhaps "consists in" is still less than transparently clear; at the very least, we should like to have some criterion for determining when one property consists in other properties. My proposal for understanding this notion is as follows: A property of a whole consists in the properties of its parts if it follows from those properties by logical or conceptual necessity.[52] This exploits the familiar idea that, in a valid deductive argument, there is nothing in the conclusion that is not already "contained" in the premises. So we can say that

51. Sellars, "Philosophy and the Scientific Image of Man," p. 26. Two points require mention here. First, this example shows it is not a consequence of Sellars's principle that in order for an object to be *F*, it must consist of parts which are *F*. This requirement would be satisfied in the case of the redness of the wall, but not in the case of the ladder's being a ladder. Second, it is clear that Sellars thinks of the object's identity as a ladder as determined by its physical structure. One might argue, however, that in order to be a ladder it must have a certain *history* (it must have been made *in order to be* a ladder, as opposed to originating by chance or for some other purpose) or that it must be *used* as a ladder. If these requirements do indeed obtain, then being a ladder would be a relational property involving the ladder's maker or user. For purposes of the present discussion, we shall ignore such historical and functional properties.

52. In saying this I commit myself to a notion of conceptual necessity (or analyticity) that is narrower than metaphysical necessity (and unlike the latter, determinable a priori) but does not require deducibility in a particular formalism, such as first-order logic. For a good recent defense of this notion, see Alan Sidelle, *Necessity, Essence, and Individuation* (Ithaca: Cornell University Press, 1989), especially chap. 5. (Sidelle's rebuttal of the arguments against analyticity in this chapter is independent of, though certainly consonant with, his endorsement of a conventionalist account of modality.)

> If an object O is a system made up of elements $e_1, e_2, e_3, \ldots e_n$, then all
> the properties of O are logical or conceptual consequences of the
> properties of, and relations between, the e_i.[53]

Thus, the fact that the visible surface of a wall is made up of red bricks implies that the wall itself is red. (The color of an opaque object is the color of its visible surface; the visible surface of the wall is identical with the visible surfaces of the bricks that make it up; all the visible surfaces of the bricks are red; therefore, the wall is red.) And from the fact that an object consists of such-and-such bits of wood in such-and-such relations, it follows that the object is a ladder; one might say that it follows from the description that the object "fits the definition" of a ladder.[54] On the other hand, it does *not* follow from the microphysical properties of an ice cube that the cube is pink,[55] and so pinkness is *not* a property of an ice cube considered as a system of subatomic particles.[56]

53. This applies only to the *monadic* properties of O. For relational properties of O, one must consider the "system" made up of O *and the other relata*.

54. Admittedly, the task of giving a definition of the physical structure of a ladder—i.e., of stating the logically necessary conditions for an object to be a ladder—is not a trivial one. And any such definition will be infected with vagueness; there is not, for instance, a sharp line of demarcation separating ladders from stairways. But these problems do not invalidate the point being made here.

55. One can, to be sure, give a "physical definition" of pink and other colors in terms of wavelengths of light. But such definitions would not capture the *experiential quality* associated with "pink"; thus they are not definitions of color terms as we ordinarily use them.

An interesting application of the principle of reducibility has been suggested to me by Philip Quinn. Water is, pretty clearly, a "system of objects," namely hydrogen and oxygen atoms. And water has the property of being liquid under standard conditions of temperature and pressure. But the predicate "x is a liquid" belongs to thermodynamics and does not occur in atomic theory; thus one cannot deduce, from the properties of hydrogen and oxygen atoms as described in atomic theory, that water is a liquid. So the principle of reducibility seems to be too strong.

What is at issue, however, is not what is logically derivable within a particular theory, but what is implied by the properties that hydrogen and oxygen atoms *actually possess*. There is no reason whatever to doubt that the characteristics in virtue of which water is a liquid "consist of" various properties and relations of the water molecules (e.g., the lack of rigid bonding between molecules), and that a material whose molecules are characterized by these properties and relations necessarily behaves in such a way that it will be classified as a liquid.

56. But could we not replace the principle of reducibility, as here proposed, with the weaker requirement that properties of systems should be *supervenient upon* the

I think it is clear that a computer is, in Sellars's phrase, "in a strict sense, a system of objects." It is the very fact that the computer's computational capabilities are deducible from, and thus explicable in terms of, the properties and relations of the computer's parts that enables us to design computers and to understand their functioning—and this, in turn, is just what makes the computer attractive as a model for the understanding of mind. But the unity-of-consciousness argument places an important barrier in the way of this project, by pointing to a property of the mind—my awareness of my present visual field—which is *not* a logical consequence of the properties of and relations between the brain's physical parts.[57]

We may conclude, then, that the mind is *logically irreducible* to the brain, in that it has properties which are not logically implied by the properties of, and relations between, the physical parts of the brain. But from this it follows, in view of the principle of reducibility, that the mind is also *ontologically irreducible* to the brain; in Sellars's phrase, the mind is "correlated with" rather than "consists of" the brain's parts. This does not,

---

properties of their parts? Nothing that has been said so far is inconsistent with the possibility that mental properties may supervene upon physical properties. But if the supervenient property is "being aware of the visual field," the question to be asked is, What is this a property *of*? If it is a property of the biological organism, then exactly the same questions arise as have already been discussed—so property supervenience has no effect on the argument.

57. It might be suggested that the property of being aware of one's visual field is in fact identical with some complex physical property, and that this complex property *is* logically implied by the properties of the parts of the brain. This raises the question of criteria for property identity; if (as seems plausible) property identity requires logical equivalence, then mental properties in general will not be equivalent to physical properties. But suppose the requirement is weaker—say, metaphysical equivalence, coextensiveness in all possible worlds? There are of course objections (e.g., "multiple realizability") to supposing that even this weaker relationship holds in general between mental and physical properties. For present purposes, however, I make no use of such objections. But I do have a specific objection in the case of "being aware of one's visual field" and similar properties, as follows: Being aware of one's visual field is essentially a unitary property, one which necessarily does not consist of and is not implied by properties of parts of a person each of which does not have the property in question. Any complex physical property which might be invoked in this context, on the other hand, would be essentially such that it necessarily consists of and is implied by the properties of the parts of the object that possesses it. So the mental property necessarily lacks a (second-order) property which the physical property necessarily possesses, and so they cannot be the same property.

to be sure, go very far toward telling us, in a positive sense, what the mind *is* or how its existence is to be explained. But it does tell us something about what the mind is *not* and about how its existence *cannot* be explained, and for a topic as contentious as this one that is a major gain.

It would perhaps be generally conceded that, *as things now stand*, we do not see how such properties as "being aware of one's visual field" could be deduced from properties of, and relations between, the neurons in the brain. But is there any reason to believe this state of affairs is permanent? Is this not merely one more case in which we can confidently wait for the advance of science to relieve difficulties which at a given juncture seem insurmountable?[58]

It is always difficult to resist such appeals to the future progress of science, in view of the astonishing progress which has already been made. Nevertheless, resistance may sometimes be called for. The difficulty in the present case is not merely that the particles of physics lack certain properties (e.g., color) that they apparently would need in order for the properties we find in the world to be deduced from them. What is at issue is that, whatever properties we may encounter in the world, these properties are co-presented for us *in a unified experience*—or, we may also say, *to a single subject*. The notion that such an experience may in the final analysis *not* be the experience of a single subject—that in fact the experience inheres in a number of different entities, each of which does *not* have that experience as a whole—is I think simply unintelligible.[59]

58. Thus, Allison alleges that Kant's is "an argument from ignorance, in the sense that it appeals to an excessively narrow, essentially eighteenth century conception of scientific explanation" (Allison, "Kant's Refutation of Materialism," p. 18; cf. also Wilson, "Leibniz and Materialism," p. 512).

59. For a discussion of Sellars's own way of dealing with this problem, see "Emergentism," pp. 484–86. Additional light is thrown on the unity-of-consciousness argument by the following quotations from Brentano and Chisholm. First, Brentano:

When someone thinks of and desires something, or when he thinks of several objects at the same time, he is conscious not only of different activities, but also of their simultaneity. . . . Now if we find the perception of seeing in one thing and the perception of hearing in another, in which of these things do we find the perception of their simultaneity? Obviously, in neither of them. It is clear, rather, that the inner cognition of one and the inner cognition of the other must belong to the same real unity (Franz Brentano, *Psychology from an Empirical Standpoint* [London: Routledge and Kegan Paul, 1973], p. 160, quoted in Roderick Chisholm, *The First Person: An Essay on Reference and Intentionality* [Minneapolis: University of Minnesota Press, 1981], p. 87).

The materialist who has followed the discussion to this point has in effect three options. She may take the heroic path laid out for her by eliminativism, by simply denying that it is ever true that a person is aware of a visual field (or of any other complex fact). Or, she may become a logical reductionist, asserting that such complex mental properties are after all logically derivable from the physical properties of the brain's parts. For many, however, the most appealing course may be to reject the principle of reducibility, enabling us to attribute to the brain as a whole (or to the person as a whole) properties which are not implied by the properties of and relations between its constituents. In order to explore this possibility, we will next consider an extremely interesting discussion of panpsychism initiated by Thomas Nagel.

Nagel and Panpsychism

In his paper entitled "Panpsychism,"[60] Nagel argues for that position on the basis of four premises, "each of which is more plausible than its denial, though perhaps not more plausible than the denial of panpsychism."[61] The four premises are:

1. *Material composition*: Living organisms are complex material systems, made exclusively of ordinary matter.
2. *Nonreductionism*: Mental properties are not physical properties of the organism[62] and are not implied by physical properties alone.

Chisholm then comments as follows:

> Brentano is telling us simply that, when a person is aware that he is seeing something and also aware that he is hearing something, then he is also aware that he is both hearing something and seeing something. Could we perhaps settle for less—say that the person who sees something and hears something is aware that the seeing and the hearing are 'parts of the same consciousness' or that they are 'compresent in consciousness'? I think not. What could it mean to say they are 'parts of the same consciousness' or 'co-present in consciousness' other than that the same person is aware of both? (Chisholm, *The First Person*, p. 88)

60. In Thomas Nagel, *Mortal Questions* (New York: Cambridge, 1979), pp. 181–95.
61. Ibid., p. 181.
62. Here I do not think Nagel is in contradiction with Searle's contention that mental properties *are* physical properties. In Searle's usage, mental properties are

3. *Realism*: Mental properties really are properties of the organism (i.e., eliminativism is false).
4. *Nonemergence*: "There are no truly emergent properties of complex systems. All properties of a complex system that are not relations between it and something else derive from the properties of its constituents and their effects on each other when so combined."[63]

Nagel then goes on to say:

> Panpsychism seems to follow from these four premises. If the mental properties of an organism are not implied by any physical properties but must derive from properties of the organism's constituents, then those constituents must have nonphysical properties from which the appearance of mental properties follows when the combination is of the right kind. Since any matter can compose an organism, all matter must have these properties. And since the same matter can be made into different types of organisms with different types of mental life (of which we have encountered only a tiny sample), it must have properties that imply the appearance of different mental phenomena when the matter is combined in different ways. This would amount to a kind of mental chemistry.[64]

For our purposes the crucial premises of Nagel's argument are Nonreductionism and Nonemergence. Unfortunately, Nagel's explanations of these premises (especially Nonemergence) are not as clear as one might wish, so we shall need to sharpen the focus. Furthermore, the interpretations of the two premises must be related in a certain way if the argument is to succeed. Nonemergence states that the properties of wholes *must* be related to the properties of parts in a certain way, and Nonreductionism states that the mental properties of persons *cannot* be derived from the physical properties of matter. So for the argument to work, the sense in which mental properties *can't* be derived from physical properties has to

---

physical properties merely in virtue of being properties of a physical organism. Nagel, on the other hand, is denying that they are physical properties in a sense "roughly equivalent to Feigl's 'physical$_2$' " (ibid., p. 183n).

63. Ibid., pp. 181–82. I have deviated from Nagel's wording in some cases; he gives a short paragraph to explain each of the premises. But I am confident that these formulations capture his meaning.

64. Ibid., p. 182.

be the same as the sense in which mental properties *have to be* accounted for by *some* properties of matter, thus forcing the ascription of mental (or proto-mental) properties to the ultimate constituents of matter.

According to Nonreductionism, mental properties are not physical properties and are not implied by physical properties alone. It seems most natural and plausible to take "implied by" here as representing logical implication, and we shall so understand it. But if so, then Nonemergence must be understood as requiring that the properties of a whole are logically implied by the properties of its parts. According to James Van Cleve, however, this

is setting the standards for explainability extremely high. If a property of a whole that follows only with nomological necessity from the properties of its parts is held on that score to be *inexplicable*, won't we have to say the same about any effect that does not follow with logical necessity from its cause? If so, the momentum of Nagel's argument will be such that it can't be stopped short of *complete causal rationalism*—Spinoza's view that causes and effects are related by logical necessity. This is a view that few philosophers nowadays would embrace.[65]

Van Cleve concedes, however, that there is a way to defend Nagel's premise short of this. If we distinguish between determinative relations that hold *at* a time and those that hold *over* time, then it may be more plausible to insist on the relation of logical implication between the properties of a whole and those of its parts for synchronic than for diachronic relations. "The argument against causal rationalism is at its strongest in the case of effects that succeed their causes in time; for how can what happens at one moment have any logically necessary connection with what happens at any other?"[66] Nevertheless, Van Cleve finds even this weaker requirement (which he terms "mereological rationalism") to be implausibly strong; he holds that "there is no reason why emergent properties can't follow with *nomological* necessity from properties of the parts, and in that case they would not be inexplicable for anyone but a causal rationalist."[67]

Now mereological rationalism, as defined by Van Cleve, is equivalent

65. James Van Cleve, "Mind-Dust or Magic? Panpsychism versus Emergence," *Philosophical Perspectives* 4 (1990): 217.
66. Ibid., p. 218.
67. Ibid.

to the principle of reducibility discussed earlier in this paper. So it becomes necessary for me to explain why the requirement of logical reducibility, as stated by that principle, is the correct one, and should not be replaced by Van Cleve's weaker requirement of nomological reducibility.

Causes may in some sense necessitate their effects, but the connection does not seem to be one of logical or conceptual necessity; thus causal rationalism is implausible. But the principle of reducibility does not require causal rationalism. Suppose that in a complex whole there are two properties, occurring simultaneously, which are linked by nomological necessity. Note that the principle of reducibility does not, so far, require a logical connection between these properties. The connection between the two properties can be as opaque, conceptually speaking, as you please. However, if one of the two properties is said to be a property of the *system as a whole*, the principle poses a question: what exactly is it that *has* the property? By hypothesis, the whole simply *is* "in the strict sense, a system of objects"; there *is no* whole "over and above" the parts of which it is composed. So whatever nonrelational properties the whole has must consist of properties of, and relations between, the parts; there simply is nothing else of which they *could* consist. If a property of the whole is not logically grounded in the properties of the parts, then it is "floating in mid-air," unattached to any real individual—but this is unintelligible.

To see this more clearly, we briefly recapitulate the main argument. A person is aware of some complex object—say, a complicated visual field. By hypothesis, it is the person's body—specifically, her brain—that is aware of this object. Now, take the smallest region of the brain that contains all the information for the visual field. This brain-region is a whole consisting of parts, but none of its proper parts can be aware of the visual field as a whole. And the properties of, and relations between, the parts do not logically imply that there is an awareness of the visual field.

So far, Van Cleve would agree, but he urges that "there is no reason why emergent properties can't follow with *nomological* necessity from the properties of the parts." Now if one state of affairs follows from another with nomological necessity, it is because there is a *causal process* linking those two states of affairs. Causal processes, furthermore, occur in and/or between concrete individual objects. So what concrete objects are available for the causal processes that result in awareness of the visual field? There are the parts of the brain region in question, of course—but the properties of, and relations between, the parts are already accounted for, and by hypothesis they do *not* imply that there is an awareness of the visual field. And the brain as a whole is not an *additional* concrete object over and above its parts, any more than, in Gilbert Ryle's example, Oxford University is an additional object over and above the colleges, libraries, and so on of which it consists. So Van Cleve's causal processes either occur in and/or between the parts of the brain, in which case they do *not* imply awareness of the visual field, or they do not occur at all.

Since this argument is crucial for the chapter as a whole, it may be helpful to state it here in a semiformalized version, as follows: Let 'p' designate a human being, by hypothesis a physical object consisting of parts $e_1, e_2, \ldots e_n$. 'Ap' says that p is aware of her visual field. 'R' is a complex relational predicate such that '$Re_1, e_2, \ldots e_n$' describes all the properties of and relations between the parts of p. Ap is the state of affairs such that 'Ap' is true. 'Imply' stands for conceptual implication. And now the argument:

1. $Re_1, e_2, \ldots e_n$, and Ap, and $Re_1, e_2, \ldots e_n$ does not imply Ap.
2. $Re_1, e_2, \ldots e_n$ causally necessitates Ap.   Provisional Assumption
3. There is a causal process (CP) connecting $Re_1, e_2, \ldots e_n$ with Ap.
      From 2
4. CP consists of events involving concrete individuals.
5. The concrete individuals involved in CP include at most $e_1, e_2, \ldots e_n$ and p.
6. p is not a concrete individual distinct from $e_1, e_2, \ldots e_n$.
7. The concrete individuals involved in CP include at most $e_1, e_2, \ldots e_n$.
      From 5, 6
8. All the events involving $e_1, e_2, \ldots e_n$ are already included in $Re_1, e_2, \ldots e_n$.
9. CP is not a causal process that connects $Re_1, e_2, \ldots e_n$ with Ap.
      From 1, 6, 7, 8
10. It is not the case that $Re_1, e_2, \ldots e_n$ causally necessitates Ap.
      From 2–9, indirect proof

I conclude, then, that the principle of reducibility (a.k.a. mereological rationalism) is correct, and under this interpretation Nagel's argument appears to be successful. Ironically, however, the very principle that undergirds the argument also defeats the conclusion he seems to want to draw from it. Recall the reference to "mental chemistry," whereby matter has "properties that imply the appearance of different mental phenomena when the matter is combined in different ways." Nagel seems to assume that, once we attribute mental (or proto-mental) properties to the ultimate constituents of matter, it may indeed be possible to explain the mental properties of human beings (such as being aware of one's visual field) in terms of their parts in a way that satisfies the principle of reducibility (= Nonemergence = mereological rationalism). But it's clear that this simply cannot be done; to repeat it once more, a person's being aware of a complex object cannot consist of parts of the person being aware of parts of the object. So the result of Nagel's argument is indeed ironic: it defeats not only the materialist views he rejects but also the "atomistic panpsychism" he intends it to support.

Interlude: Van Inwagen on Thinking

Some of Peter van Inwagen's thoughts shed an interesting light on the unity-of-consciousness argument. In his book *Material Beings*,[68] van Inwagen argues that most ordinary material objects, such as rocks, mountains, and stars, automobiles and computers, simply do not exist. He further, and surprisingly, contends that this radical thesis does not contradict ordinary beliefs about these matters. The propositions we ordinarily assert using sentences apparently referring to such objects can still be regarded as true—or, alternatively, as falsehoods that for most practical purposes can be treated as true.[69] For, in the case of a chair, the "chair-receptacle" (viz., the region of space that we would ordinarily say is occupied by a chair) is in fact occupied by a large very number of physical simples; these simples are "arranged chairwise," and through their collective interactions they perform all the functions we normally associate with chairs—they remain rigid under stress, support weight against the action of gravity, reflect light at certain wavelengths so as to create a characteristic visual impression, and so on. There is no danger, then, that when one goes to sit down one will land on the floor as a result of van Inwagen's metaphysical revisionism!

At this point one might be inclined to ask, *what is the difference* between a collection of simples arranged chairwise (as van Inwagen would have it) and a chair? The answer to this question is straightforward: the

68. Ithaca: Cornell University Press, 1990. Section 12, which is central for the present discussion, is entitled "Unity and Thinking."

69. Ibid., pp. 102–3. (In this respect, he argues, his views on this topic parallel the views commonly taken about assertions such as "the sun went behind the tree," which apparently presuppose pre-Copernican astronomy.)

simples in question do not *compose* anything; there *is no* whole of which they are parts. Quite simply, there are no chairs.

Van Inwagen's reasons for this intriguing thesis lie beyond the scope of the present inquiry. What is germane, however, are his reasons for the sole exception to this ban on composite objects: namely, organisms. There really are organisms; the simples of which they are made really are parts that compose a whole. So the inventory of the physical universe, at its most general level, is as follows: there are simples, and there are organisms composed of simples. There is nothing else.

But why must we say that organisms, unlike other apparent composite objects, really do exist? The answer is that some organisms think.[70] And thinking, unlike the sorts of things done by chairs and other nonorganisms, is not the kind of thing that can be done cooperatively by a collection of simples. Van Inwagen puts it like this: "The simples that are arranged shelfwise cooperate to support weight; the simples that are arranged sidereally cooperate to produce light. Our initial impression is that there is a certain huge object, the sun, that does a thing called shining. Later, under the influence of our theory, we decide that what we took to be the product of the activity of a single object was the product of the joint activity of many."[71] But couldn't this be true of thinking also? Could we not

> say that one's thinking is really no more than a persistent habit of cooperation among certain simples? And if we have the cooperating simples, what need have we for *one*?
>
> In my view, we do have a need for "one," that is, for the individual thing that thinks. I do not see how we can regard thinking as a mere cooperative activity. Things can work together to produce light. They may do this by composing a single object—a firefly, say—that emits light. But things that work together to produce light are not forced, by the very nature of the task set them, to produce light *by* composing a single object that emits light. . . . But things cannot work together to think—or, at least, things can work together to think only in the sense that they can compose, in the strict and mereological understanding of the word, an object that thinks.[72]

70. Van Inwagen notes that he is "using 'think' in a very liberal sense, sufficiently liberal that I will count such items as *feeling pain* as instances of thinking" (ibid., p. 118).

71. Ibid.

72. Ibid. This point enables van Inwagen to make a quick and elegant disposition

Van Inwagen concludes, "I therefore exist. And yet I have parts. . . . Therefore, there is at least one case in which a material being has parts."[73]

He goes on to explain that, while thinking shows the necessity of recognizing the existence of composite material objects, it does not explain or account for the fact that it is possible for there to be such objects. (It is because the composite objects exist that they are able to think, not the other way around.) Considering further his own case (which is, of course, the source of the reflections concerning thinking), van Inwagen says, "it seems to me to be plausible to say that what binds [the simples that compose me into a single being] is that their activities constitute a life, a homeodynamic storm of simples, a self-maintaining, well-individuated, jealous event."[74] And if the simples whose activities constitute van Inwagen's life are thereby bound into a single being, one must, in order to avoid being arbitrary, say the same concerning other human beings, other animals, and indeed concerning organisms generally. Organisms, then, really do exist, unlike all other sorts of composite material objects.

By this time it is hoped that the reader will have become aware of a certain irony. Van Inwagen has argued, on the basis of the claim that thinking cannot be the composite activity of a collection of objects, that a human being—the thing that thinks—must exist as a whole composed of physical parts. In this chapter I have argued, starting from essentially the same premise, that the thing that thinks *cannot* be a whole composed of physical parts. How is this difference to be accounted for? Assuming neither of us has been simply careless, it would seem there must be some deep philosophical difference between us that accounts for the diametrically opposed conclusions we derive from the same starting point. One candidate for this difference might be van Inwagen's materialism concerning human beings, which is rather an assumption than a conclusion of the present argument. But this still leaves the question unanswered: Why does van Inwagen fail to appreciate [what I claim to be] the antimaterialistic implications of the "unity of thinking"?

The most obvious answer to this further question is that van Inwagen

---

of the problem of artificial intelligence: "computers—computers of the sort IBM sells—cannot think. They cannot think because they do not exist" (ibid., p. 119).

73. Ibid., pp. 119–20.

74. Ibid., p. 121. For further explanation of the terms employed, see van Inwagen's excellent discussion of the nature of organisms on pp. 81–97.

rejects the Principle of Reducibility. That is to say: he does not explicitly reject this principle, since he never explicitly considers it. But it does seem that he is *committed to* rejecting it; it seems he is committed to affirming that the "system of objects" in question—namely, the human being—has properties that do not "consist of properties of, and relations between, its constituents." For if thinking *could* consist of properties of, and relations between, the simples arranged humanwise, van Inwagen's argument for the existence of human beings as composite physical objects would collapse. And if the Principle of Reducibility is false, then thinking might indeed be the activity of a human being considered as a composite physical object.

Nevertheless, a further irony remains. The claim that organisms think, as described above, would seem to require that "the properties of organisms are not wholly determined by, do not wholly supervene upon, the properties of their parts." The thesis that this is so, according to van Inwagen,

> is sometimes called holism. According to holism, even a complete and correct list of principles of composition would not enable a perfect reckoner—the Laplacian Intelligence, say—to reckon the properties of wholes from the complete truth about the intrinsic properties of and the relations that hold among the parts that compose the wholes. Whether holism is correct, I do not know. Like most of my contemporaries, I am strongly inclined to think it is not correct, though I can't put my finger on what my reasons for thinking so are.[75]

But the inference seems inescapable: If holism is false, then a "perfect reckoner" could indeed "reckon the properties of wholes [in this case, of the persons that think] from the complete truth about the intrinsic properties of and the relations that hold among the parts that compose the wholes." If this were possible in the case of thinking, then van Inwagen's argument for the existence of persons (rather than mere collections of simples) would be fallacious. But van Inwagen holds, in agreement with the argument of this chapter, that the property of thinking *cannot* be "a mere cooperative activity" of various simples. It follows from this that either holism is true, and complex material objects can have properties that "are not wholly determined by, do not wholly supervene upon, the prop-

75. Ibid., p. 90.

erties of their parts," or else the subject of thinking is something other than the material composite. No holism, then, means no unity of thinking for a composite material organism. I believe we are forced to conclude that van Inwagen's reflections on this topic are a work still in progress.[76]

Process Philosophy and the Self as Continuant

As an argument for dualism, the unity-of-consciousness argument so far is in a certain way incomplete. One's consciousness of a complex state of affairs, we have argued, is the state of a single, unitary subject; it can't be "parceled out" to a number of distinct individuals. But the principle of reducibility is, as we've seen, synchronic rather than diachronic, so it can't be used to establish the identity and continuity of the self over time. Doesn't this leave open the possibility that what we have, instead of a single individual existing over a period of time, is rather a *series* of such individuals—in the terminology of process philosophy, a "serially ordered society of actual occasions"? Note that this does not amount to the "bundle theory" of the self embraced by Hume among others. The bundle theory is vulnerable to the unity-of-consciousness argument, because it fails to clearly account for the co-presence at the same instant of a number of diverse types of experiences in a single, unified consciousness. But the "actual occasions" of process thought are, in effect, momentary selves, momentary persons—and the hypothesis is, that a person in the ordinary sense is a series of such momentary selves rather than a single, continuing individual.[77]

To be sure, one's present experiences often include memories of past experiences, and memories of them *as one's own experiences*, as experiences of the very same self that is now recalling them in the present moment. As I remember starting to write this chapter, for instance, my memory is that *I myself*, the very same individual now writing these words, some time ago started with the section on Swinburne. Emphatically, I do *not* recall that some other individual, related to me by some complicated causal-histori-

76. I must disagree, then, with the remark that immediately follows the passage last quoted: "Fortunately, none of the questions I attempt to settle in this book will require a decision about the correctness of holism" (p. 90).

77. According to Whitehead, "An enduring personality in the temporal world is a route of occasions in which the successors with some peculiar completeness sum up their predecessors" (Alfred North Whitehead, *Process and Reality* [New York: Macmillan, 1929], p. 531).

cal connection, began the writing of the chapter. Still, all this depends on memory, and it must be conceded that memory is more subject to mistake than one's awareness of one's present experiences. Does this fallibility of memory provide an opening for the self-as-series theory to establish itself as a plausible alternative?

It is, no doubt, in some sense conceivable that what I recall as my own past experiences are not really my own experiences but rather those of other individuals who stand in some special relation to me. It is also conceivable that none of the things I remember ever happened at all—that they are all pseudo-memories implanted in me by a malignant demon or a devious brain researcher. But is there any reason to take these "skeptical doubts" seriously as real possibilities? Lacking any such reason (and I know of none), I submit that the proper attitude to such doubts is simply to ignore them, to operate (as we always in fact do operate) on the basis of the Principle of Credulity, which says that, in the absence of any reason to think otherwise, we ought to assume that the way things seem to us to be is the way things are.

It may, however, be possible to press the argument beyond this point. It does seem to be a matter of present experience that I as a person not only *perceive* but also *act*—that the *very same individual* who perceives that the peach is ripe reaches up his hand to pick it off the tree. Now even the simplest actions take more than the tenth of a second or so during which "actual occasions" are supposed to endure; furthermore, humans are capable of forming and carrying out plans over periods lasting weeks, months, or years. This fact serves to bring out a paradox in the process theory of actual occasions. Clearly, the process of "concrescence," in which the occasion selects its "subjective aim" and is guided thereby in the way it incorporates past experiences into the present, is modeled on the process of decision making by actual persons as it is empirically observed. Yet the time frame for an actual occasion (which is never actually specified but can hardly be more than a small fraction of a second) is far too brief for any meaningful decision making to occur. And what are we to make, on this scheme, of the important phenomenon of deferred gratification? Actual occasions, we are told, seek to "maximize enjoyment," but in cases of deferred gratification the enjoyment may only be experienced by a self many generations removed from the one that decides to accept present pain for a better ultimate result. A truly hedonistic actual occasion, it would seem, ought to "live for the moment" in the strictest sense of the term! In view of all this, I submit, the only reasonable stance to take is that the consciousness that experiences diverse phenomena at a single moment also endures through time—that the very same self now typing these words is the one that several years ago conceived the plan for this book.

In retrospect, it can now be seen that the arguments of the last three chapters tend to converge around a single conclusion. In Chapter 3 it was argued that rationality requires a responsiveness to normative and teleological considerations and that this responsiveness must somehow pertain to the human person as a whole, since we cannot reasonably suppose the behavior of elementary particles to be influenced directly by norms and objectives. In Chapter 4 it was argued that the choices in virtue of which we are agents must also be ascribed to the person in a holistic sense; to analyze them as resultants of a congeries of events on the microlevel is in effect to make the choices disappear. And in the present chapter we have argued that the self which is the subject of conscious experience must be a unity of a sort that is inconsistent with its being a whole consisting of physical parts. But what *is* this self? What sort of metaphysical account can be given of its nature? That is the question we must now address.

Problematic Dualisms

According to the argument of the preceding chapter, the subject of awareness must be a continuing individual not composed of physical parts. If that is correct, then (provided we are realists about the physical) some form of substance dualism is inescapable. But the most common forms of dualism are subject to serious objections, as well as some that are not so serious. We begin in this chapter with critiques of Cartesian dualism (with attention to a variant form proposed by Richard Swinburne) and Thomistic dualism, before proceeding in the next chapter to a version of dualism that may prove to be more viable.

CARTESIAN DUALISM

Cartesian dualism will be taken as comprising the following claims: Human bodies are ordinary physical things, made up of ordinary matter in complicated arrangements, and bearing roughly the properties attributed to such bodies by contemporary physics. Bodies, however, have no mental or psychological properties whatever. Human minds, or souls, are completely nonphysical; they possess no mass, extension, or location. These souls are the ultimate subjects of all mental and psychological properties; they sense colors, feel pains and pleasures, entertain plans, and decide upon courses of action. Under normal circumstances each soul causally interacts with a particular body from which it receives sensory information, and through which it executes its intentions. Bodies, of course, reproduce through the normal biological processes, but souls do not; instead, they are individually created by God. As simple substances, they are incapable of dissolution into parts and so possess a natural immortality.

Such is the dualism bequeathed to us by Descartes, and in its main outlines it survives in the writings of contemporary dualists.[1] Most readily dispensable, perhaps, is the claim about the soul's natural immortality; since Kant, philosophical proofs of immortality have been suspect, and contemporary dualists are generally satisfied to point out that dualism is logically compatible with the belief in a future life. The notion that souls are individually and specially created by God is not always insisted on, but it is probably the only viable answer to the question about the origin of souls; views in which souls originate biologically hardly qualify as Cartesian.

OBJECTIONS TO CARTESIAN DUALISM

It is common today in some circles to object to mind-body dualism for broadly ethical reasons. Carol Zaleski states the matter thus: "One often hears it said that dualism is the besetting vice of Western thought. We have inherited a life-denying, world-rejecting dichotomous way of thinking, we are told; the language of soul-body dualism, in particular, is blamed for everything from the despoiling of the environment to the oppression of women."[2] It is difficult to know how seriously to take this sort of charge. To ignore the complaints may seem disrespectful of those who hold such opinions. A lengthy discussion, on the other hand, runs the risk of ascribing to the charges an intellectual weight they do not possess. To be sure, many of those in Western culture who have been especially guilty of despoiling the environment and subjugating women may have been

1. For instance, John Foster, *The Immaterial Self: A Defense of the Cartesian Dualist Conception of the Mind* (London: Routledge, 1991). Interestingly, Foster himself accepts a form of idealism, but in this book he assumes provisionally a Cartesian dualist view of the body.

2. Carol Zaleski, *The Life of the World to Come: Near-Death Experiences and Christian Hope* (New York: Oxford University Press, 1996), p. 54. Zaleski goes on to say, "What this critique overlooks is the great variety of dualisms."

dualists. After all, most Westerners who have held any views at all on the mind-body issue have been dualists. But those who suppose that the despoliation of the environment and the denigration of women are unique to the West are strangely blind to the lessons of history. The suspicion that dualism fosters an alienation of the person from her body and a low valuation of the body is somewhat more plausible. Yet these attitudes are not an inevitable consequence of metaphysical mind-body dualism, which is perfectly compatible with the biblical view that the human body, and the material world generally, are the good creations of God.

A philosopher who has taken particular care to respond to these charges is Charles Taliaferro.[3] In doing this Taliaferro has developed what he terms an "integrative dualism" in which "persons are integrally related to their bodies so that the person and his or her body function as a singular unit mentally and physically. The person and body are not, strictly speaking, metaphysically identical. They are separable individuals, but all this is in keeping with the proximate, materially conditioned, embodied nature of personal life."[4] Elaborating on the integrative nature of personal embodiment, Taliaferro writes:

> What it means for a person to be embodied is for a person to bear some of the following relations to his or her body, relations I characterize in the first person. I am sensorily bound up with my body in that I feel with these fingers and this skin, I see with these eyes, smell with this nose, eat and taste with this mouth, hear with these ears. I can have some sensory apprehension of the inside of my body, under the skin so to speak, as when I feel my heartbeat or have a stomach ache, feel thrills and internal pain. . . . It is by means of my being embodied that I have the conscious life I have; it is by means of my body as a whole that I interact with other persons and material objects. In my acting in the world, my body is transparent to my intentions in the sense that . . . I can . . . act in a basic, nonmediated fashion. . . . I act, but not in virtue of performing some other action.[5]

I believe it is clear that any dualism that is viable in the present intellectual climate must be an integrative dualism in Taliaferro's sense. Against such a

3. See Charles Taliaferro, *Consciousness and the Mind of God* (Cambridge: Cambridge University Press, 1994), especially pp. 114–22, 226–47.
4. Ibid., p. 115.
5. Ibid., pp. 116–17.

dualism, the charges cited above are ineffective. Furthermore, all of the major types of dualism to be discussed are at least logically consistent with integrative dualism, though some types may harmonize with it more readily than others. The charges that have been noted serve as reminders to avoid distorted versions of dualism as well as ways of speaking about dualism that suggest (perhaps unintentionally) these distorted versions; they do not provide us with a reason to reject dualism as such.

The hoariest objection specifically to Cartesian dualism (but one still frequently taken as decisive) is that, because of the great disparity between mental and physical substances, causal interaction between them is unintelligible and impossible.[6] This argument may well hold the all-time record for overrated objections to major philosophical positions. What is true about it is that we lack any intuitive understanding of the causal relationship between Cartesian souls and bodies. And there is no doubt that, other things being equal, a mind-body theory that allowed such understanding would seem preferable to one that did not. The reason this is not decisive is that, as Hume pointed out, *all* causal relationships involving physical objects are at bottom conceptually opaque.[7] We find the kinetic theory of gases, with its ping-pong-ball molecules bouncing off each other, fairly readily understandable. This, however, is only because we have learned from experience about the behavior of actual ping-pong balls, and our expectations in such cases have become so habitual that they seem "natural" to us; we have no ultimate insight into the causal relations involved, except to say, "That's the way things are." But equally, and emphatically, the "way things are" includes the facts that our thoughts, feelings, and intentions are influenced by what happens to our bodies, and vice versa; to deny these palpable facts for the sake of a philosophical theory seems a strange aberration.

An interesting variant of the argument against the possibility of mind-body interaction comes from Brian Leftow:

> Bodies act only by sending physical signals or exerting physical energy. But a physical signal can only arrive at some place in space. Again, if physical energy reached an unlocated soul, it would disappear from physical space.

6. Some additional philosophical objections to dualism are discussed in Chapter 8 below.
7. This may not be true, however, of mental causation. There does seem to be something transparently intelligible about the fact that desiring a certain state of affairs tends to make one act so as to bring it about.

This would violate the conservation laws. . . . Again, if an unlocated soul causes motion in a body, it causes the body to have a certain amount of kinetic energy. As the soul causes this, the energy has no physical source; it is physically *ex nihilo*. If the soul does not transfer the energy to the body, the soul brings it into existence without transferring it from *any* source. . . . [This] is equivalent to bringing a certain amount of mass into existence physically *ex nihilo*. . . . Theists might justly protest that this is God's specialty, and He does not share it.[8]

This ingenious argument begs the question by assuming that psychophysical causation must work in the same way as physical causation. Why can't the soul register what is going on in the brain without absorbing energy from the brain? (What would the soul do with that energy, anyway?) And can't there be two energetically equivalent brain-states which encode different "instructions" from the soul? If not, why not? The dualist besieged by arguments of this sort would do well to repeat to herself, from time to time, "Psychophysical causation is *not* physical causation."[9]

Still another variant surfaces in Jaegwon Kim's "pairing problem" for mind-body dualism.[10] In its simplest form, the problem is to specify what it is that connects a Cartesian soul with a particular body so as to enable causal interaction between them. With ordinary physical causation, spatial relations play a crucial role in determining which causal connections are possible, but these relations are not available for Cartesian souls. So what, if anything, takes their place? This is certainly not a crushing objection to Cartesian dualism. At a minimum, the Cartesian can simply say that God, in creating an individual soul, establishes a unique tie between it and a certain body that enables the causal relationship between them.[11] Nevertheless, the objection does bring into vivid relief the metaphysical disparity between Cartesian souls and the bodies with which they interact.

In the light of all this, we can say that if Cartesian dualism were other-

8. Brian Leftow, review of *Philosophical Perspectives 5: Philosophy of Religion*, ed. James Tomberlin, *Faith and Philosophy* 13 (April 1996): 273.

9. For additional discussions of this objection, see Stewart Goetz, "Dualism, Causation, and Supervenience," *Faith and Philosophy* 11, no. 1 (1984): 92–108; and Keith Yandell, "A Defense of Dualism," *Faith and Philosophy* 12, no. 4 (1985): 551–53.

10. Jaegwon Kim, "Causality and Dualism" (lecture delivered at the University of Notre Dame, March 7, 1998).

11. This possibility was suggested in discussion by John Foster.

wise acceptable, the dualist would be well advised to stand her ground in spite of the objection concerning mind-body interaction. But there are other difficulties, which to me at least seem a good deal more formidable. One group of objections stems from the observable continuity (in spite of all differences) between human beings and other forms of life on earth. What are we to say of the souls of beasts? Descartes, of course, denied that they have any, and so was committed to his notorious (and monumentally implausible) doctrine of animal automatism. But his conclusion on this point, however misguided, was by no means unmotivated. As he said, if animals "could think as we do, they would have an immortal soul as well as we, which is not likely, because there is no reason for believing it of some animals without believing it of all, and there are many of them too imperfect to make it possible to believe it of them, such as oysters, sponges, etc."[12] If in spite of this we persist in supposing that at least some animals enjoy mental lives, we shall have to suppose them endowed with souls in the same sense in which humans are so endowed (though not necessarily equal to human souls in their powers). And this in turn leads into a whole tangle of intriguing questions. The doctrine of the special creation of individual souls will then have to be extended from the human case to all other ensouled creatures, a consequence Swinburne explicitly accepts.[13] This may not seem terribly problematic as applied to dolphins and horses, but it boggles the mind when we come to slugs, termites, and mosquitoes. (Swinburne apparently would limit souls to some of the "higher" types of animals.)[14] And what of Descartes's biggest concern, the doctrine of natural immortality, applied to this teeming multitude of creatures? Are we to suppose that God by fiat annihilates the souls of those animals not destined for eternal life? It does seem that some response is due from dualists on these issues, in contrast with the profound

12. Descartes to the marquis of Newcastle, in David M. Rosenthal, ed., *Materialism and the Mind-Body Problem* (Englewood Cliffs, N.J.: Prentice-Hall, 1971), p. 21.

13. See Richard Swinburne, *The Evolution of the Soul* (Oxford, Clarendon, 1986), p. 199. Foster, on the other hand, writes, "I am not entirely sure that non-human animals *do* have minds" (*Immaterial Self*, p. 237), though he admits that in everyday life he takes it for granted that they do. And the neuroscientist John Eccles, who holds to the special creation of individual human souls, maintained a stubborn agnosticism about animal consciousness throughout his lengthy dialogues with Karl Popper. See *The Self and Its Brain: An Argument for Interactionism* (New York: Springer-Verlag, 1977), pt. 3.

14. Swinburne, *Evolution of the Soul*, pp. 180–83.

silence which (with few exceptions) has prevailed up until now. Still more difficult, however, is the problem that arises with cloning, or with other situations in which an animal is physically divided into parts, with each part subsequently developing into a complete organism. Prior to the division there is one animal, and one soul; afterwards, there are two animals, and two souls—but where does the additional soul come from? Are we to suppose that God is, so to speak, on call to provide an additional soul whenever one is required?[15] And, finally, there is the difficulty that there does not seem to be any plausible or natural way to incorporate Cartesian souls into the process of biological evolution. Should we suppose that, once an organism has physically evolved, God creates for it an appropriately different kind of soul than that which animated its forbears? Or are the souls somehow implicated in the causation of the evolutionary process? One awaits with keen interest the elaboration by dualists of suitable answers to these and similar questions![16]

Another group of difficulties for Cartesian dualism results from the facts concerning the dependence of the mind on the brain and its states. To be sure, such dependence is not always a problem for dualism; to some extent it is to be expected, as can be seen in the following quotation from Swinburne:

> A person has a body if there is a chunk of matter through which he makes a difference to the material world, and through which he acquires true beliefs about that world. Those persons who are men have bodies because stimuli landing on their eyes or ears give them true beliefs about the world, which they would not otherwise have, and they make differences to the world by moving arms and legs, lips and fingers. Our bodies are the vehi-

15. This particular difficulty would look somewhat different if we were to view the origin of souls in line with the occasionalist model, as Brian Leftow has suggested is the case for Aquinas (in comments delivered at the University of Notre Dame, March 5, 1998). For a bit more on occasionalism, see the next chapter.

16. Some limited answers are given by Swinburne in *The Evolution of the Soul*, pp. 183–96. It should be noted, however, that there is a certain irony in Swinburne's title, inasmuch as the soul is precisely what does *not*, on his view, evolve. Note also that Alvin Plantinga appeals to the special creation of human souls (which he apparently accepts) in arguing that on the assumption of Christian theism the "antecedent probability" of the theory of common ancestry (that is, its probability before the specific evidence for evolution has been considered) is not especially high (Plantinga, "Science: Augustinian or Duhemian?" *Faith and Philosophy* 13 [July 1996], p. 387).

cles of our knowledge and operation. The 'linking' of body and soul consists in there being a body which is related to the soul in this way.[17]

According to this, embodiment as such requires that minds depend on brains in at least two ways: they need the brains as sources of information about the environment, and also as channels through which to issue instructions for bodily action. So it is readily understandable that brain misfunction (as the result, for example, of a blow to the head) should deprive the mind of sensory information and render voluntary bodily action impossible.

But the observed dependence of mind on brain goes far beyond this. On the dualistic view, why should *consciousness itself* be interrupted by drugs, or a blow on the head, or the need for sleep?[18] And why should reasoning, generally thought of as the distinctive activity of the conscious mind, be interrupted by such physical disturbances? The natural conclusion from Cartesian dualism would seem to be that consciousness should continue unabated during such times—deprived, to be sure, of sensory input and the capacity for motor action. There are, furthermore, the well-publicized facts about the effects of mind-altering drugs, as well as the apparently permanent changes in personality and character that can result from brain injuries. By making the mind essentially independent of the brain rather than dependent on it, Cartesianism deprives itself of a ready explanation for these kinds of dependence that we actually find.

An especially striking example of dependence is found in the phenomenon of *visual agnosia*.[19] In this condition, persons who have suffered damage to a specific portion of the brain become impaired in their ability to process visual information. These individuals suffer no loss of visual

17. Swinburne, *Evolution of the Soul*, p. 146. Clearly an integrative dualism such as Taliaferro's would insist on a much richer account than this of the linking of body and soul. The challenge, however, is to provide a plausible and compelling rationale for this closer linkage within the framework of Cartesian dualism.

18. I am not here insisting on the point that since, according to Descartes, consciousness (or "thinking") is the essential property of the soul, it is metaphysically impossible for the soul to exist without being conscious. No doubt the defining property can be taken to be the potentiality for consciousness, rather than consciousness as such. The problem remains of giving a plausible account, within the Cartesian scheme, of the kinds of dependence we actually find.

19. See Neil R. Carlson, *Physiology of Behavior*, 5th ed. (Boston: Allyn and Bacon, 1994), pp. 171–80.

acuity, nor is their general intelligence impaired. But they lose the ability to "make sense" of what they are seeing in ways we ordinarily take for granted. One sufferer from this syndrome found himself unable, after a motorcycle accident, to identify individual faces, or to "read" the expressions on people's faces—obviously a serious disadvantage for normal social interactions. The point is this: what we have here, as a result of specific damage to the brain, is a disruption not of sensory capacity as such but rather of an extremely subtle and sophisticated type of information processing; just the sort of thing one would expect Cartesianism to assign to the conscious mind rather than to the brain.

Or is this what we should expect of the Cartesian? Philip Quinn doubts this: "I'd be willing to bet that we will learn from neuroscience, if we have not already done so, that this processing of visual information goes on in the brain. And if a Cartesian mind requires as input processed rather than unprocessed visual information in order to perform such tasks as identifying faces and reading facial expressions, it will come as no surprise that it can't perform such tasks if its damaged brain can't provide it with the processed visual information it requires as input."[20]

The problem with this lies in the strong probability that, as neuroscience progresses, more and more of our "advanced" mental processes will be found to be associated with, and dependent upon, specific brain processes.[21] When that happens, Quinn's strategy would lead us to view the conscious mind as essentially a passive spectator, enjoying awareness of the results of the various mental processes but contributing little or nothing of its own to those results. And this is an outcome no Cartesian dualist can possibly welcome.

Charles Taliaferro takes a slightly different tack: "Allow for interaction, and I see no reason why we shouldn't expect the connection to be intricate and many-layered, replete with information processing."[22] This modifies the Cartesian scheme in a different way than Quinn's suggestion: rather than merely increasing the intellectual work assigned to the brain at the expense of the conscious mind, Taliaferro suggests a complicated, multi-

20. Comments delivered at the Midwestern Metaphysics Workshop, University of Notre Dame, August 14, 1997.

21. Visual agnosia is by no means an isolated phenomenon. For an instructive overview of the ways in which various cognitive functions can be disrupted by brain disorders, see D. Frank Benson, *The Neurology of Thinking* (New York: Oxford University Press, 1994).

22. Personal correspondence.

layered system with mind and brain interacting on a number of different levels. This does indeed present a more "integrative" picture of the mind-brain relationship. But the picture is not one that readily harmonizes with Cartesianism. The defining property of mind is the potentiality (at least) for conscious experience; matter, on the other hand, entirely lacks any such potentiality. Why, then, do we find that, under the conditions of embodiment, the mind requires the assistance of all this cerebral apparatus in order to perform its inherent function? Does this mean that in a disembodied state, without the brain's assistance, the mind is unable to think? Some contemporary dualists might be willing to accept this, but it is hardly in the spirit of Cartesianism. (For one thing, it effectively undermines Descartes's modal argument for dualism, which many contemporary dualists still want to use. Descartes, in Meditation 2, found himself able to conceive not only the *existence* of the soul without any body but its *functioning* in that state exactly as it does in everyday life; thus his inability to be sure that this is not in fact the way things are. So by the principle, "the conceivable is the possible," it should follow that it is *actually possible* for the mind to function lacking a body.)[23] Or is it that the mind can function perfectly well without the brain while disembodied, but embodiment itself creates the need for the cooperation of the "meat computer" in order for the mind to work properly?[24] The supposition that this

23. Might the Cartesian dualist object that it is merely a contingent fact that the mind is unable to function without a brain, a fact that is not inconsistent with the metaphysical possibility of its doing so? I don't think so—the inability in question would be the consequence of the mind's *essential nature*, and thus would be necessary rather than contingent. Some dualists might suggest that when we conceive of the mind functioning without the brain we are (unknown to ourselves) conceiving of its functioning thus with the special assistance of God, which somehow compensates for the lack of a brain. But many of us find that (like Descartes in Meditation 2) we can conceive of the mind's functioning disembodied without any such special support. This could perhaps be shown *not* to be consistently conceivable if there were a successful version of the ontological argument, one which shows (as Plantinga's version admittedly does not) that it is impossible consistently to conceive of the nonexistence of God. So the modal argument for dualism would be viable only for successful ontological arguers—a conclusion that might not distress Descartes, but would be uncomfortable for a number of contemporary dualists!

24. It is interesting in this connection that W. D. Hart, like Taliaferro an advocate of Descartes's modal argument for dualism, has stated in discussion that in his view the argument shows that the mind is in fact able to perform essentially all of its functions while disembodied.

may be the case is nicely illustrated by the following passage from Richard Purtill:

> As to the way in which damage to the brain inhibits thoughts, the non-materialist might give the following analogy: as long as I am dressed in fairly close-fitting garments, my garments move when my body moves. But it is the body which normally moves the clothing, not vice versa. Still, someone can make my arm move by grabbing my sleeve and pulling on it. And if my sleeve gets somehow immobilized, e.g., by getting caught in some machinery, it can prevent my arm from moving. Suppose I fell into a vat of quick-drying cement and my clothing got soaked with it and then the cement hardened, so I was imprisoned in a rigid set of garments. I couldn't move so long as I stayed in those garments. But if I could get free of those garments, I could again move normally either without garments or in a new set of garments.[25]

In that case, embodiment—in the "close-fitting garment" of mortal flesh—would constitute a *burden and limitation* on the natural powers of the soul. But one would hardly expect this idea to be welcomed by an integrative dualist such as Taliaferro!

It seems, then, that the actual dependence of mind and personality on brain goes far beyond what one would naturally expect on the basis of Cartesian dualism. It's not that these phenomena are logically inconsistent with Cartesianism; no doubt they can be accommodated, but there is a price to be paid for doing so. If Cartesian dualism is to be taken seriously as the truth about minds and their bodies, then both the problem of dependence and the problem of continuity need to be addressed in a way that exhibits the known facts as plausible consequences of the underlying metaphysical view. A string of ad hoc conjectures will not suffice.

THE SWINBURNE VARIATIONS

Swinburne has already been cited several times as an exemplar of Cartesian dualism, and in general this seems the right way to classify his

---

25. My thanks to Richard Purtill for providing me with a copy of some work in progress, from which this quotation is taken. I should add that Purtill, whose own view of the mind-body relationship is broadly Aristotelian, does not himself endorse the rather "Platonistic" approach implied in the quotation.

views. But there are some special features of those views that demand separate notice, features that are not readily harmonized with his overall
Cartesian stance. One surprise comes in his suggestion that under certain
conditions there is a *conventional element* in our affirmation of the existence of souls. He asks whether "the soul is still there when the man is
asleep, having no conscious episodes," and replies, "This calls for a decision of what (if anything) we are to mean by saying of some soul that it
exists but is not functioning."[26] If we say that a person still exists during
sleep while having no conscious life, what we mean, according to Swinburne, is that the sleeping body will again by normal processes give rise to
a conscious life which by normal criteria of personal identity will be the
life of the person existing prior to the loss of consciousness. This fact,
however, could be described in either of two ways. "We could say that
souls exist only while conscious; while a person is asleep, his soul ceases to
exist but it is made to exist again when he is woken up. But this would be
a cumbersome way of talking. It is better to understand by a soul existing
when not functioning that normal bodily processes on their own will, or
available artificial techniques can, make that soul function. In saying this
I am laying down rules for the use of a technical term, 'soul'."[27] No doubt
most of us will agree with Swinburne's linguistic preference here. But
what is striking is his suggestion that whether or not souls exist during un-
consciousness is a matter of convention. It is really puzzling to see how
this view can be combined with Swinburne's creationism: How can it be a
matter of convention whether God has created a soul once, or many
times?

Swinburne says other things that seem to call his creationism into
question. Apropos of artificially synthesized animals, he writes, "There is
no reason to suppose that souls will come into existence only through the
normal sexual processes. Laboratory synthesis which produces the same
physical organism should produce the same mental life." He also says, "It
is possible that a new person is produced within the same body as the former person when the *corpus callosum*, the main nerve-link between the
two cerebral hemispheres is severed."[28] But on the creationist view neither
sexual processes, nor laboratory synthesis, nor brain surgery can possibly

26. Swinburne, *Evolution of the Soul*, pp. 176–77.
27. Ibid., p. 177. (Swinburne states that he sees nothing contradictory in saying
that a substance has many beginnings of existence.)
28. Ibid., p. 197.

produce a soul; the creation of souls is the prerogative of God alone. It is possible, to be sure, that these quotations should be understood charitably, as a concession to the reader who has not yet become aware of Swinburne's creationism. (Creationism is first explicitly introduced in the section just following the two quotations.) But the shift is hardly innocuous. It is one thing to suggest that commissurotomy, by dividing the brain, may produce a second mind, and to add that additional scientific research will be needed to answer this question. It is quite another matter to propose that scientific research will enable us to tell whether, on the occasion of a brain operation, God in his infinite wisdom chooses to create a new person!

Further perplexities arise in the course of Swinburne's discussion of mind-brain interaction. He assumes that it would be possible for scientists to compile a long list of correlations between brain-events and conscious experiences, stating what kind of experience occurs when a given sort of brain-event occurs. But he doubts that scientists could ever progress beyond this to a proper scientific theory, in which a relatively small number of laws would explain all the observed correlations. The most fundamental reason for this is, that

> Brain-events are such different things qualitatively from pains, smells, and tastes that a natural connection between them seems almost impossible. For how could brain-states vary except in their chemical composition and the speed and direction of their electrochemical interactions, and how could there be a natural connection between variations in these respects and variations in the kind of respects in which tastes differ—say, the differences between a taste of pineapple, a taste of roast beef, and a taste of chocolate—as well as the respects in which tastes differ from smells and smells from visual sensations? There does not seem the beginning of a prospect of a simple scientific theory of this kind.[29]

In view of the apparent impossibility of producing a scientific explanation of brain-mind correlations, Swinburne proposes that they should instead be given a personal explanation, in terms of the actions and purposes of God.[30] As to the way in which God does this, "the suggestion is that God

29. Ibid., pp. 189–90.
30. This argument is already given in Swinburne's *The Existence of God* (Oxford: Clarendon, 1979), chap. 9. A somewhat similar argument is developed by Robert M.

has given to each animal brain and each human brain a limited nature, as it were; a limited nature such that in the circumstances of normal embodiment it keeps a soul functioning in predictable ways—without the brain having that nature deriving from any general law of brain/mind connection, stating in general under what circumstances souls function and have the particular mental events they do."[31]

This suggestion raises fascinating, not to say baffling, questions. Apparently the "limited nature"[32] given to the brain consists in new causal powers which enable it to keep the soul functioning. Here is one question: Does God confer those new causal powers by endowing the elementary constituents of the brain—the quarks, gluons, and electrons—with causal powers they would not otherwise possess? Or are the new powers somehow given to the brain as a whole, without any change in the natures of the constituent particles? If the latter is the case, then we have here a holistic metaphysic of causation that surely needs a great deal more explanation—an explanation Swinburne has not provided. If on the other hand the change occurs in the elementary particles, what happens when those particles leave the brain and others take their place? For that matter, why couldn't God confer the enhanced natures on all elementary particles without exception, so that when they are appropriately combined they give rise to a conscious life with no further direct divine intervention? But, finally, isn't there some kind of redundancy in invoking both the divine special creation of a soul and the conferral by God of a special nature on the body so as to enable it to interact with the soul? When creating the soul, why doesn't God just give it the power to interact with matter, and have done?

I shall not attempt to explain in detail the reasons for these deviations by Swinburne from the expected Cartesian positions. In the interest of consistency, he would be better off to toe the Cartesian line more strictly, and give up talk of the soul's existence as conventional, of physical events as producing souls, and of God's endowing brains with new natures they would not otherwise possess. Yet there is this point to notice: It seems to

---

Adams in "Colors, Flavors, and God," chap. 16 in *The Virtue of Faith* (New York: Oxford University Press, 1987).

31. Swinburne, *Evolution of the Soul*, pp. 198–99.

32. I must confess that I fail to understand the sense in which these natures are "limited," unless this means that the power to sustain a mental life is effective only under a limited range of circumstances.

me that all of these deviations result, in different ways, from Swinburne's attempt to do more justice to a scientific perspective on the mind and brain than pure Cartesianism readily allows for. In each case, furthermore, the apparent inconsistency arises because of the conflict with these ways of speaking of Swinburne's creationism. I suggest, then, that if he were to do away with the creationism, the way might open up to an approach that would in the end be more satisfactory than the one Swinburne gives us. It's clear, however, that Swinburne isn't willing to take that step, and so we have to leave the tensions in his view unresolved.

THOMISTIC DUALISM

The difficulties with Cartesian dualism fall into a consistent pattern; they can all be seen as resulting from an excessive separation between souls and the natural world—both the soul's own body and the broader world of nature of which that body is a part. Both the general difficulty of mind-body interaction and more specific difficulties about the extent of interaction fit this pattern, as do the problems about the relationships between the human soul and the rest of nature. This theory has in it too much of Plato, one might say, and not enough of Aristotle.

But if this is a correct diagnosis, it also suggests a possible solution. In the thought of Thomas Aquinas, we have available an Aristotelian dualism that is more developed than Aristotle's own theory, and one that establishes the closer, one might say friendlier, relations between mind and nature whose lack creates such difficulties for Descartes. This theory provides an account of the lower forms of life, expresses the continuity between these forms and the human race (humans share the "vegetative soul" with all living things, and the "sensitive soul" with the animals), yet stresses the uniqueness of human beings as rational, moral, and above all immortal creatures. Indeed, Aquinas's theory has been seen as a sort of halfway house between dualism and materialism, or as a demonstration that the dichotomy between dualism and materialism is a mistake.[33] Clearly, this is an option we need to explore.

Rather than going direct to Aquinas's text, we will be examining some contemporary versions of Thomistic anthropology; this will enable us to

---

33. See Eleonore Stump, "Non-Cartesian Substance Dualism and Materialism without Reductionism," *Faith and Philosophy* 12 (October 1995): 531.

circumvent some perplexities of textual interpretation and also to view Thomistic dualism in a form that is more directly relevant to the current debate. We will be relying primarily on an exposition by Eleonore Stump, with some reference also to the work of David Braine.[34]

Central to the thought of both Aristotle and Aquinas on this topic is the assertion that the soul is the form of the body. Accordingly, Stump begins her exposition with an account of Aquinas's doctrine of "substantial forms," the "forms in virtue of which a material composite is a member of the species to which it belongs" (p. 508). As a contemporary illustration for this doctrine, she discusses in some detail a particular protein, designated as C/EBP, which is known to play a role in regulating gene expression. She summarizes Aquinas's doctrine thus:

> In general, the form (the substantial and accidental forms taken together) of a material object is the arrangement of the matter of that object in such a way that it constitutes that object rather than some other one. This arrangement will commonly be a function not only of the shape of the matter, but also of the properties of the material parts and the ways those parts relate causally to each other. Form for Aquinas is not static but dynamic, something that includes the functioning of and the causal interactions among the parts. (p. 509)

For this reason, the form of a living body is destroyed when death occurs, even though the general shape and arrangement of parts may remain the same for a considerable time; death marks the end of the "functioning of and the causal interactions among the parts" of the living body.

A characteristic feature of Aquinas's (and Aristotle's) doctrine is that a material object typically appears as a sort of structured hierarchy of matter and form: what at one level is analyzed as a matter-form composite becomes in turn "matter" for the form at the next higher level of organization. Thus, in the case of the C/EBP protein, protons, neutrons, and electrons are configured into atoms, which in turn are configured into amino acids, and so on until we arrive at the extremely complex structure of the protein itself. Nevertheless, it is Aquinas's view that "a given material object has only one substantial form." In terms of the example, "there is just one substantial form for C/EBP which makes it the kind of molecule it is,

---

34. Ibid., pp. 505–31; David Braine, *The Human Person: Animal and Spirit* (Notre Dame, Ind.: University of Notre Dame Press, 1992), especially pp. 499–531. (Page references in the text of this section are to Stump's article.)

a dimer of amino acid subunits composed of certain elements related to one another in particular ways and themselves comprised of other components ordered in particular ways." One might say, then, that the form of the substance as a whole "subsumes" the forms of the components of the substance, in such a way that the forms of the components no longer exist as such. Applying this to the human case, "When Aquinas says that the soul is the form of the body, he means that it is the single, substantial form of the body" (p. 509).

Summarizing her survey of Aquinas's doctrine of form, Stump asserts "it seems not unreasonable to think that by 'form' Aquinas means an essentially configurational state" (p. 509). Two additional points deserve mention. In virtue of the forms that constitute them, physical objects often display what Searle calls "emergent$_1$ properties."[35] Such a property is defined by Stump as "a feature or property of a whole or system [that] is not a property of the parts of that system, and can be explained in terms of the properties of the parts of the system and the causal interactions among those parts" (p. 510).[36] Secondly, the forms of individual substances, according to Aquinas, are not universals but rather particulars; in this respect they seem similar to the "tropes" or "abstract particulars" featured in some contemporary ontologies.[37]

35. See John Searle, *The Rediscovery of the Mind* (Cambridge: MIT Press, 1992), pp. 111–12. (For more discussion of Searle's conception of emergence, see the next chapter.)

36. Stump goes on to suggest that the C/EBP protein may exhibit features that are, in Searle's sense, "emergent$_2$"; that is, features that "can't be explained just in terms of the properties of the individual parts of the system and the causal interactions among those parts" (p. 510). This seems to be a mistake. Her basis for saying this is that the protein's ability to regulate genes depends on its shape, which in turn may depend, not merely on the properties and causal interactions among the atoms that constitute the molecule, but on "the interactions of the atoms of the molecule with enzymes that catalyze folding" (p. 510). But the fact that an enzyme is involved in producing the protein's final shape doesn't mean the shape is emergent$_2$, any more than the fact that a watch must be assembled by a watchmaker makes its shape and function emergent$_2$. In both cases, ordinary causal explanation works well enough, and there is no need for this "more adventurous" sense of emergence. (Searle doesn't think *anything* is emergent$_2$; he remarks that "the existence of any such features would seem to violate even the weakest principle of the transitivity of causation" [Searle, *Rediscovery of the Mind* p. 112].)

37. See David Lewis, *On The Plurality of Worlds* (New York: Oxford, 1986); and Keith Campbell, "The Metaphysics of Abstract Particulars," *Midwest Studies in Philosophy* 6 (1981). As a substance-attribute ontologist, Thomas would not, of course, accept D. C. Williams's view that tropes are the fundamental elements of which every-

Turning now to the human soul in particular, we note that it is called the "intellective soul" or "rational soul," in virtue of the distinctive sorts of activities that are possible for it; nevertheless, this soul is "that configuration on the basis of which something exists as this living human body" (p. 511). So there is not in the human being a "vegetative soul" responsible for nutrition, growth, and reproduction, and an "animal soul" responsible for sensation. Rather, it is in virtue of this one form that is the rational soul that "a human being exists as an actual being, as a material object, as a living thing, as an animal, and as a human being with cognitive capacities" (p. 511). Contrary to Descartes, Thomas holds that the soul has a location: "while the body is alive, the soul is located where the body is" (p. 512). Again contrary to Descartes, the human being is not identical with her soul; he says that "a human being is not a soul only but rather a composite of soul and body."[38]

Nevertheless, the individual that is a human being is able to continue in existence after the body has perished, even though the individual in that disembodied state is not a complete human being. And the account given so far makes this possibility of disembodied existence difficult to account for. How can the form of the C/EBP protein continue to exist after its constituent atoms have been recycled? The obvious answer is that it can't.

So an explanation is needed here, and Thomas's expositors do their best to provide it. David Braine addresses the criticism (by Anthony Flew) that Thomas is guilty of a "category-mistake" in treating the abstract notion of form as if it were the notion of a concrete being. In reply, Braine insists that "the notion of 'form' as the correlate of matter is not a determinate one—and, what is important here, it is not even determinate as to category."[39] Because this is so, " 'Form' may either refer to accidental relations, or to non-subsistent principles of unity and operation of non-accidental wholes (syllables, molecules, and living things provide very different cases of this), or, if the human soul be subsistent, the subsistent principle of unity and operation in the human being."[40] The principal reason given by Aquinas for supposing that the human soul is subsistent [i.e., concrete] is that it has operations proper to it—namely, those of rational thought—

---

thing else in the world is constituted and that concrete particulars are collections of tropes.

38. *Summa Theologiae* I, q75 a4.
39. Braine, *Human Person*, pp. 510–11.
40. Ibid., p. 511.

that have no bodily organ. Surprisingly, Braine agrees with this claim of Aquinas, and provides elaborate (but unconvincing) arguments in its favor,[41] blithely disregarding the scientific evidence to the contrary. (Stump, on the other hand, acknowledges that Aquinas was mistaken on this point; in so doing she deprives herself of Aquinas's major philosophical reason for the subsistence of the soul.)

Eleonore Stump arrives at a similar conclusion through a different route. She notes the objection, considered by Aquinas himself, that "forms dependent on matter as regards being don't have being themselves, strictly speaking; rather, the composites have being through the forms."[42] Her response is that "these perplexities stem, at least in part, from too limited a view of what Aquinas has in mind with the notion of form" (p. 513). She goes on to correct this "too limited view," and her remarks about this need to be quoted at some length.

> On Aquinas's view, to be is to be configured or to have a form, and everything is what it is in virtue of a form. . . . [F]or Aquinas the ability of matter to be configured is just a consequence of the fact that matter has being, and what is fundamentally configured is not matter but being. . . . An angel, for Aquinas, is immaterial but configured since it has order and species, that is, since it is a kind of thing with one rather than another set of characteristics. . . . Understanding this point helps to explain why although Aquinas is perfectly content to deny matter of God, he refuses to deny form of God: being, even divine being, is configured.
>
> This broader understanding of form is useful for the problems under consideration, *because it introduces an ambiguity into the notion of form.* There are forms, such as the form of C/EBP, which are forms in the sense that they give a configuration to something. And then there are forms that don't configure something else but that are rather themselves configured. On Aquinas's view angels are forms in the latter sense. (pp. 513–14; emphasis added)

Furthermore,

> it is helpful to recognize that it is possible for something to be both configured and a configurer of other things. This is so, in familiar and unprob-

41. See Braine, *Human Person*, chap. 12, "The Second Refutation of Mechanism: Linguistic Understanding and Thinking Have No Bodily Organ," pp. 447–79.
42. *Summa contra Gentiles* II, 51.

lematic ways, as regards material objects. So, for example, C/EBP is configured. . . . But it is also a configurer. When it is bound in the right way to DNA, it helps to unravel the DNA molecule, thereby reconfiguring the DNA in such a way as to make transcription possible. (p. 514)

And as regards the human soul,

For Aquinas, the metaphysical world is ordered in such a way that at the top of the metaphysical hierarchy there are forms—God and the angels— which are configured but which aren't configurational constituents of anything else. . . . Near the bottom of the hierarchy are forms that configure matter but don't exist as configured things in their own right. . . . And in the middle are human souls, the amphibians of this metaphysical world, occupying a niche in both the material and the spiritual realm. Like the angels, the human soul is itself configured; but like the forms of other material things, the human soul has the ability to configure matter. The human soul, then, is a configured configurer. (pp. 514–15)

The lucidity and elegance of this exposition shouldn't cause us to overlook the admission that the initial problem is resolved only by rendering the term 'form' ambiguous! But given these explanations, Stump's fundamental account of Aquinas's doctrine is in place; she goes on to refine some details and to place Aquinas in the context of our contemporary discussion. In a brief summary statement, she asserts that "the soul is an essentially configurational state which is immaterial and subsistent, able to exist on its own apart from the body" (p. 519).

Stump is readily able to establish that Aquinas's doctrine is a form of non-Cartesian substance dualism, though one reservation is required: the soul, for Aquinas, is not itself a complete substance (though he does call it a substance), but rather a constituent of the psychophysical whole which is the human being. But Stump has more difficulty in establishing that his doctrine can be viewed legitimately as some form of materialism. In the process of arguing for this, she makes several additional points that distinguish Aquinas's dualism from that of Descartes. Perhaps most striking is her denial of causal interaction between soul and body: "As a part of this [metaphysical] sort, the form couldn't interact causally with the matter it informs. The form has causal influence in the sense that the composite has the causal influence it does because of the form. But it makes no sense to think, for example, of the configuration of C/EBP interacting causally with the matter of C/EBP" (p. 518). She then proceeds to establish some points of contact between Aquinas and contemporary materialists. Citing Patricia Churchland's assertion that a main characteristic of physicalism is that "mental states are implemented in neural stuff,"[43] she claims that Aquinas is a physicalist in this sense, because "although Aquinas mistakenly supposes that the intellect is tied to no particular bodily organ, he nonetheless holds that the intellectual soul is the form constituting the human body as a whole. On his view, therefore, mental states will be implemented in matter" (pp. 520–21). She also cites the materialist Richard Boyd as holding that mental states are "purely configurational," with no essential compositional properties, and she claims that Aquinas agrees with this also (pp. 521–22). Her final conclusion is that "to make progress on a philosophical understanding of the nature of the mind . . . it would be good to break down the dichotomy between materialism and dualism that takes them to be incompatible positions" (p. 523).

CRITIQUE OF THOMISTIC DUALISM

The objectives of Thomistic dualism are admirable: to preserve the doctrine of eternal life, to overcome the difficulties of Cartesianism, and to establish a closer connection between the soul, its body, and the world of nature. And given the apparently intractable opposition between materialism and dualism, the aim of transcending the dichotomy between them has considerable appeal.

But we have to determine the success of this undertaking. And the place to begin is with the notion of "form." As we've seen, both Braine and Stump admit, in their different ways, that the concept of form employed by Aquinas is ambiguous; in particular, it is ambiguous as between signifying form as abstract or as concrete. Now, there is no doubt that the word 'form' in English, and its equivalents in Greek and Latin, is ambiguous in this way as well as in others. But it is one thing to admit this fact and quite another to incorporate into one's technical philosophical terminology an admittedly ambiguous term, while doing nothing to remove the ambiguity. To do the latter is virtually a recipe for philosophical confusion; there is just too much danger that the ambiguities that have been

---

43. Cited from Patricia Churchland, *Neurophilosophy: Toward a Unified Science of the Mind/Brain* (Cambridge: MIT Press, 1990), p. 352.

tolerated in the technical term will reemerge, undetected, in the course of specific philosophical discussions.[44]

Unfortunately, this is just what happens in Stump's discussion. In setting out Aquinas's doctrine, she repeatedly appeals to considerations that are correct as regards her initial explanation of forms as abstract particulars, but which fail when applied to the notion of the soul as a "configured configurer." Consider her summary description of the soul as "an essentially configurational state which is immaterial and subsistent" (p. 519). A "configurational state" must be a state *of something*; God and the angels are not "states of" anything except themselves—and neither is the human soul. Or take the claim that "mental states are implemented in neural stuff" (p. 520). The talk of "implementation" here derives from the world of computers: programs, abstract sequences of instructions, are "implemented in" computer hardware. For Aquinas, however, mental states are implemented in the brain *together with* the immaterial, subsistent mind. Patricia Churchland, we may surmise, would not be amused! Again, consider the claim that mental states are purely configurational, with no essential compositional properties (p. 522). This may or may not be true of mental states as such, but it assuredly is not true (for Aquinas) of the mental states *of human beings*, which must of necessity be states of the human soul, and normally also of the human body. And finally, consider her claim that the form which is the soul cannot interact causally with the matter it informs, on the grounds that "it makes no sense to think . . . of the configuration of C/EBP interacting causally with the matter of C/EBP" (p. 518). Here Stump seems unaccountably to have forgotten that her final analogue for the soul is not "the configuration of C/EBP" but rather the C/EBP protein itself, which is "both configured and a configurer of other things" (p. 514). When this protein "helps to unravel the

---

44. It is noteworthy that Braine, in general a staunch defender of Aquinas, nevertheless observes that he "may have been deceived by confusing two different uses of the world 'form' of quite different origins: on the one hand, we have the use of the world 'form' to refer to "natures" or "predicates" concretely (the idioms 'The horse is a quadruped' and 'The circle is a figure' are representative here) from which there arises the notion of self-subsistent natures, within which there is no individuation correlated with matter but identity between the nature as subject of concrete predicates and the individual realizing this nature, as supposedly in the case of God and the different species of angels; and, on the other hand, we have this notion of 'form' as a correlate of matter originating with the idea of shapes, such as circularity, as forms of material things such as metal rings" (p. 499n).

DNA molecule, thereby reconfiguring the DNA in such a way as to make transcription possible," there is no way that it will not be exerting a causal influence on the DNA. Nor is there any way the soul, as conceived by Aquinas, will not be exerting causal influence on the brain and body it "configures." The Thomistic dualist has by no means succeeded in freeing herself from the problem—to the extent that it is a problem—of mind-body interaction.

Upon closer examination of Thomas's view, yet another of the problems with Cartesian dualism surfaces once again, albeit in an altered form. Consider the implications of the fact that it is only human souls which are subsisting immaterial individuals; animal souls, in contrast, are indeed "configurational states" of the matter that makes up the animals' bodies. This already puts in place a pretty wide separation between humans and the rest of animate nature.[45] And it has another implication that many dualists will find disturbing. If the apparently rich mental and emotional lives of dogs, dolphins, and chimpanzees can be fully explained in terms of the functioning of the "organized matter" of their bodies,[46] where is the plausibility of arguing that the cognitive activity of human beings requires an immaterial soul? Especially when the principal argument for such an immaterial soul has rested on the contention, now scientifically discredited, that there is no neural correlate for the higher rational processes?

This is not merely a problem for the dualist polemic against materialism; it raises serious questions about the internal coherence of Thomistic dualism. Consider the account that is to be given of sense perception for humans and other animals. In the case of animals, the subject of perception is the organized matter of the brain and nervous system. For humans, the subject is the composite consisting of the brain and nervous system *and the immaterial soul.* This contravenes what seems to be strong evidence that perception works in very much the same way in humans and in animals. And it means that the metaphysical analysis of perception in the two cases is going to have to be radically different, in spite of the empirical similarities. Unless and until these contrasting metaphysical analy-

45. Braine addresses this issue when he considers "The Objection That This View Re-Erects a False Contrast between Human Beings and the Other Animals" (pp. 538–40); I do not believe he resolves it satisfactorily.

46. This functioning, however, would not be viewed by the Thomistic dualist as mechanistic functioning, susceptible to a reductive explanation. A good deal of Braine's book can be seen as a sustained argument against the possibility of a mechanistic understanding of life processes.

ses have been spelled out—and so far as I know, this has not been done—
we have good reason to suspect here a fundamental incoherence in
Thomistic dualism. In the end, the problem of the souls of beasts is a
bone in the throat for Thomism as it is for Cartesianism.

I close this chapter by referring to yet another saying of Stump's that I
find puzzling, yet also full of promise. In the course of charting Aquinas's
affinities with certain materialistic views, she asks: "Would Aquinas think
that the mind is identical to the brain if he had known enough neuroscience? Given what he says about the separated soul, the answer, of
course, has to be 'no.' But even if we ask about the mind before death, in
its natural, embodied state, it seems less misleading to say that he would
have thought that the mind emerges from the functioning of the brain,
since the human form on his account is dynamic rather than static" (p.
520). What intrigues me in this is the suggestion that Aquinas might have
thought that "the mind emerges from the functioning of the brain." What
does Stump mean by this? Clearly, the historical Aquinas would *not* have
thought this, given his doctrine of the special creation of individual
souls—a doctrine which Stump refers to several times. Does she mean
that he *should* have thought this way, and perhaps would have done so
had more scientific knowledge been available to him? Or is the notion of
emergence here to be understood in some way that is compatible with
Aquinas's creationism?

Whatever the answers to these questions may be, the notion that a substantial soul emerges from the functioning of the brain—a theory of sub-
stance-emergence, and not merely property-emergence—strikes me as
immensely promising. I believe that it may contain the seed of a solution
to the mind-body problem that offers more hope, and is subject to fewer
objections, than any of the views surveyed up till this point. But this
"good news" must wait to be spelled out in the next chapter; the news in
this one has been mostly bad.

CHAPTER SEVEN

Emergent Dualism

Anyone who finds traditional dualisms implausible, yet is unsatisfied with eliminativist or strongly reductive views of the mind, is likely to find congenial the idea that somehow or other the mind emerges from the functioning of the brain and nervous system. This idea was an integral part of the "emergent evolution" espoused early in this century by Samuel Alexander, C. Lloyd Morgan, and C. D. Broad.[1] The notion of emergence, however, has been employed with a variety of different meanings, so it is necessary at this point to say something about those differing uses and to clarify the way emergence is to be understood in the present discussion. We will then proceed to consider two recent emergentist views of the mind, offer a constructive proposal, and conclude by raising some additional questions.

CONCEPTS OF EMERGENCE

We begin with some intriguing (but, as we shall see, incomplete) remarks of John Searle about different concepts of emergence. He writes:

1. For a brief overview, see T. A. Goudge, "Emergent Evolutionism," in *The Encyclopedia of Philosophy*, 2:474–77. A fuller treatment is found in Brian P. McLaughlin, "The Rise and Fall of British Emergentism," in A. Beckerman, H. Flohr, and J. Kim, eds., *Emergence or Reduction? Essays on the Prospects of Nonreductive Physicalism* (Berlin: de Gruyter, 1992), pp. 49–93.

Suppose we have a system, S, made up of elements $a, b, c \ldots$ For example, S might be a stone and the elements might be molecules. In general, there will be features of S that are not, or not necessarily, features of $a, b, c \ldots$ For example, S might weigh ten pounds, but the molecules individually do not weigh ten pounds. Let us call such features "system features." The shape and the weight of the stone are system features. Some system features can be deduced or figured out or calculated from the features of $a, b, c \ldots$ just from the way these are composed and arranged (and sometimes from their relations to the rest of the environment). Examples of these would be shape, weight, and velocity. But some other system features cannot be figured out just from the composition of the elements and environmental relations; they have to be explained in terms of the causal interactions among the elements. Let's call these "causally emergent system features." Solidity, liquidity, and transparency are examples of causally emergent system features.

On these definitions, consciousness is a causally emergent property of systems. It is an emergent feature of certain systems of neurons in the same way that solidity and liquidity are emergent features of systems of molecules. The existence of consciousness can be explained by the causal interactions between elements of the brain at the micro level, but consciousness cannot itself be deduced or calculated from the sheer physical structure of the neurons without some additional account of the causal relations between them.

This conception of causal emergence, call it "emergent$_1$," has to be distinguished from a much more adventurous conception, call it "emergent$_2$." A feature F is emergent$_2$ iff F is emergent$_1$, and F has causal powers that cannot be explained by the causal interactions of $a, b, c \ldots$ If consciousness were emergent$_2$, then consciousness could cause things that could not be explained by the causal behavior of the neurons. The naive idea here is that consciousness gets squirted out by the behavior of the neurons in the brain, but once it has been squirted out, it then has a life of its own.

It should be obvious from the previous chapter that on my view consciousness is emergent$_1$ but not emergent$_2$. In fact, I cannot think of anything that is emergent$_2$, and it seems unlikely that we will be able to find any features that are emergent$_2$, because the existence of any such features would seem to violate even the weakest principle of the transitivity of causation.[2]

2. Searle, *The Rediscovery of the Mind* (Cambridge: MIT Press, 1992), pp. 111–12.

This remarkable (but somewhat less than perspicuous) passage calls for comment at several points. First, consider the system features that are explainable merely in terms of the features of the elements and the way they are arranged. Apparently Searle doesn't find it appropriate to describe these features as emergent in any sense; there just isn't enough that is "new" in the system feature to warrant such a description. We should note, however, that these features can on occasion be quite interesting and surprising. Consider, for example, the complex and beautiful properties exhibited by fractal patterns, which surely are wondrous enough in spite of being "mere" logical consequences of the equations that generate the patterns. Perhaps we can term such features (at least when they are striking enough to get our attention) as "logical emergents," or as "emergent$_0$."

To be distinguished from these are the causally emergent system features—emergent$_1$ features—which are the main topic of Searle's discussion. But before considering these there are some points about "explanation" that need to be clarified. First, I think Searle is using this notion in such a way that if A is the cause of B, then A explains B; there is no further requirement on explanation beyond what is already implied in causation. But, second, Searle clearly is talking about both causation and explanation *as they would be in a completed science.* In saying that "the existence of consciousness can be explained by the causal interactions between elements of the brain at the micro level" he clearly doesn't mean that we can *now* give such explanations; it's obvious we can't do this, since we don't yet know even the Humean-type correlations that govern the emergence of consciousness.

According to Searle, emergent$_1$ features are those which "cannot be figured out just from the composition of the elements and environmental relations; they have to be explained in terms of the causal interactions among the elements."[3] And such, Searle claims, are liquidity, solidity, transparency—and consciousness. But there is a further, important distinction that needs to be made here. In order to grasp this distinction, consider the case of biological life, which is clearly an emergent$_1$ feature. And let's consider two scenarios with respect to the explanation of life. In both scenarios, we assume that physics and chemistry, including organic chemistry, have been studied and understood to the point that we have completely adequate theories to explain and predict everything that happens in the ab-

---

3. Although he doesn't say so, I am sure Searle means to include here also possible causal interactions between the elements and the environment.

sence of biological life. In the first scenario, these theories also enable us to explain and predict what occurs on the biological level; the "causal interactions among the elements," described according to the ordinary laws of physics and chemistry, suffice to explain life as well as everything that happens on the pre-biological levels. Whether or not this is true about life, we are confident it is true of solidity and liquidity. Emergent feature of this sort we term "ordinary causal emergents," or "emergent$_{Ia}$."

But not all of Searle's emergent$_I$ features need be of this sort. In fact, it is clear that emergentists have typically wanted to assert more than this. And it is at least conceivable that, in the case of life, the most complete understanding possible of the causal interactions of nonliving matter will *not* enable us to explain life itself. On this second scenario, the processes of life would indeed be explained by causal interactions among the elements, but *the laws that govern these interactions are different because of the influence of the new property that emerges in consequence of the higher-level organization.* So if (for example) consciousness is emergent in this sense, the behavior of the physical components of the brain (neurons, and substructures within neurons) will be *different*, in virtue of the causal influence of consciousness, than it would be without this property; the ordinary causal laws that govern the operations of such structures apart from the effects of consciousness will no longer suffice. If, then, the behavior of the system continues to be law-governed, we will have to reckon with the existence of *emergent laws*, laws whose operation is discernible only in the special sorts of situations in which the higher-level emergent properties manifest themselves.[4] Features which are emergent in this more robust sense may be termed "emergent$_{Ib}$"; they involve "emergent causal powers" whose operation is described by the emergent laws.[5]

4. Presumably these laws would be formulated in such a way as to have the ordinary, non-emergent laws as a special case. The possibility of such laws is the principal theme of P. E. Meehl and Wilfrid Sellars, "The Concept of Emergence," in H. Feigl et al., eds., *Minnesota Studies in the Philosophy of Science*, vol. I (Minneapolis: University of Minnesota Press, 1956), pp. 239–52. (Meehl and Sellars are discussing a paper by Stephen Pepper in which he claims to show that emergent properties are of necessity epiphenomenal. Their reply, in effect, is that these properties will not be epiphenomenal if there are emergent laws, a possibility Pepper mistakenly dismissed.)

5. Note that while this concept of emergence is defined in relation to an ideal, completed science, our estimate concerning which (if any) properties will turn out to be emergent in this sense is likely to be influenced by the science we currently have available. Brian McLaughlin attributes the demise of classical emergentism to the rise

According to Timothy O'Connor, emergentists often have wanted to claim for the emergent properties they recognize "novel causal influence"—in effect, emergence$_{1b}$. O'Connor explains novel causal influence as follows: "This term is intended to capture a very strong sense in which an emergent's causal influence is irreducible to that of the micro-properties on which it supervenes: It bears its influence in a direct, 'downward' fashion, in contrast to the operation of a simple structural macro-property, whose causal influence occurs *via* the activity of the micro-properties that constitute it."[6] The notion of "downward" causal influence—a notion that is quite popular in recent discussions of emergence—is of course a metaphor; the "levels" involved are levels of organization and integration, and the downward influence means that *the behavior of the "lower" levels—that is, of the components of which the "higher-level" structure consists—is different than it would otherwise be, because of the influence of the new property that emerges in consequence of the higher-level organization.*[7]

This is what "downward causation" *needs* to mean, if the notion is to make coherent sense. But the terminology of levels and of downward causation has, I want to suggest, a definite tendency to mislead one into thinking of the different levels as concrete and as capable of exerting, on their own, distinct kinds of causal influence. It's as though one were thinking of a multistoried building, in which almost everything that reaches the upper stories comes in through the ground floor, but occasionally something comes down from the upper stories—say, a telephone

---

of quantum mechanics, which made it more feasible than before to derive macro-properties (including chemical and biological properties) from micro-properties (McLaughlin, "Rise and Fall of British Emergentism," pp. 54–55).

6. Timothy O'Connor, "Emergent Properties," *American Philosophical Quarterly* 31 (April 1994): 98.

7. For an argument that downward causation is ultimately an incoherent concept, see Jaegwon Kim, "Making Sense of Emergence," *Philosophical Studies* (forthcoming), and " 'Downward Causation' in Emergentism and Nonreductive Physicalism," in Beckerman, Flohr, and Kim, *Emergence or Reduction*, pp. 119–38. Kim, however, does not consider the possibility of emergent laws, which are essential to the notions of emergence$_{1b}$ and downward causation as explained here. Brian McLaughlin, on the other hand, comes very close to the idea of emergent laws in his conception of "configurational forces" (see "Rise and Fall of British Emergentism," pp. 52–53). McLaughlin considers the idea coherent, though he says "there seems not a scintilla of evidence that there is downward causation from the psychological, biological, or chemical levels" (p. 55).

call containing orders or instructions—that makes things on the ground floor go differently than they would otherwise have done. It's obvious, though, that this kind of picture is seriously misleading. The only concrete existents involved are the ultimate constituents and combinations thereof; the only causal influences are those of the ultimate constituents in their interactions with each other, and the only way the "higher levels" can make a difference is by *altering or superseding the laws* according to which the elements interact. To think otherwise is to fall into confusion on the basis of an inadequate grasp of what "levels" and "downward causation" involve.

The kind of emergence we've been considering here—emergence$_{1b}$, the emergence of novel causal powers—has been discussed by Jaegwon Kim in several essays. Kim views emergentism as, in effect, a special form of supervenience theory.[8] Even more interesting, however, is his general claim that "nonreductive physicalism . . . is best viewed as a form of emergentism."[9] Kim would reject such a view because it violates the "causal closure of the physical," a notion we've discussed at length in Chapter 3.[10] This becomes part of Kim's case against nonreductive physicalism, a view which seeks to maintain the reality and causal efficacy of mental properties, yet uphold the physicalist commitment to the causal closure of the physical realm. Kim recognizes, however, that the classical emergentists were not committed to the closure of the physical, and thus would not be deterred by this objection.[11]

There remains Searle's suggestion—and dismissal—of the "more adventurous" conception of emergence he dubs "emergent$_2$." On the one hand, his remarks about the transitivity of causation make the view seem just incoherent: if A causes B, and B causes C, then A causes C, and how could anyone deny this? And yet, the image of consciousness being "squirted out" by the behavior of the neurons, and then acquiring a "life of its own," strikes one as intelligible even if a bit bizarre. Just what is going on here?

8. See Jaegwon Kim, "The Nonreductivist's Troubles with Mental Causation," in *Supervenience and Mind: Selected Philosophical Essays* (Cambridge: Cambridge University Press, 1993), pp. 336–57.

9. Ibid., p. 344. And compare: "It is no undue exaggeration to say that we have been under the reign of emergentism since the early 1970s" (Kim, "Making Sense of Emergence").

10. Kim also suspects that the notion of downward causal influence is ultimately incoherent; see note 6 above.

11. Kim, "Nonreductivist's Troubles with Mental Causation," p.356.

In fact, Searle seems to be talking about two different things. According to his formal definition, consciousness is emergent$_2$ iff the existence of consciousness is emergent$_1$ and consciousness has causal powers that can't be explained in terms of the causal interactions of the neurons. And that does seem incoherent: if the neurons, by their causal interactions, generate consciousness, then in so doing they generate whatever causal powers consciousness may possess. But Searle goes on to say, "If consciousness were emergent$_2$, then consciousness could cause things that could not be explained by the causal behavior of the neurons." Here the problem is not that the neurons' behavior doesn't explain the *causal powers* of consciousness, but that it doesn't explain the *things that are caused* by consciousness in the exercise of those powers. And that is quite a different matter; at least it is if we don't assume that the causal operations of consciousness must be deterministic. If there is an indeterministic element in the operations of consciousness, so that consciousness can cause things it isn't caused to cause—if, for instance, consciousness should happen to be endowed with libertarian freedom—then it might very well be the case that "consciousness could cause things that could not be explained by the causal behavior of the neurons." Understood in this way, there is nothing incoherent about the idea that consciousness might possess "emergent$_2$" features. Since free will seems to be the most plausible example of an emergent$_2$ feature, we will label this sort of emergence as the "emergence of freedom."

Still further issues arise when we confront these concepts of emergence with the principle of reducibility discussed previously. What we have to ask is whether the emergent property of an object is one that can consist of properties of, and relations between, the parts of the object. In the case of what we have termed "logical emergents," there is no reason to doubt that this is the case. For causal emergents, Searle's "emergent$_1$" properties, the situation is more complex. Keep in mind that the principle of reducibility is synchronic rather than diachronic; it concerns the relationship between properties of parts and wholes at a given time, not with the relationship over time or with the genesis of the emergent properties. Understood in this way, there seems to be no reason in general why both "emergent$_{1a}$" properties, ordinary causal emergents, and "emergent$_{1b}$" properties, involving novel, emergent causal powers, might not pass muster when scrutinized in terms of the principle. However, some putative emergent properties may fail the test. Sellars would argue, for example, that there is no satisfactory way to treat phenomenal color as an emergent property. Since this is so, being colored must be a property of base-level objects. He maintains that "the being colored of colored objects (in the naive realist sense) does not consist in a relationship of non-colored parts."[12] Furthermore, "unless we introduce Cartesian minds as scientific objects, individual scientific objects cannot be meaningfully said to sense-redly. Nor can the scientific objects postulated by the theory of inorganic matter be meaningfully said to be, in a relevant sense, colored."[13] Because of this, Sellars introduces a new category of scientific objects, called "sensa," to be the subjects of the color predicates[14] which are the successors in the Scientific Image of the color predicates of the Manifest Image. Sensa, then, are *emergent individuals*, individuals whose genesis is governed by emergent laws. Such individuals are of necessity emergent$_{1b}$; the ordinary causal laws that govern the operations of neural structures in the absence of sensa will not account either for the existence of sensa or for their causal influence on the perceptual states of persons and on their resulting behavior.[15]

Finally, what of the properties Searle terms "emergent$_2$"? The most plausible example we have seen of an emergent$_2$ property is libertarian free will, and it seems clear that this cannot be a property that consists of the properties of, and relations between, the parts that make up a system of objects. If we are to include libertarian free will as an attribute of persons, it seems we shall need to recognize persons, or minds, or souls, as unitary subjects, not analyzable as complexes of parts. And if creationist versions of dualism are rejected, as the previous chapter suggests they should be, this means we shall have to acknowledge the existence of minds as *emergent individuals*—similar in this respect (but perhaps only in this respect) to Sellars's sensa.

TWO RECENT EMERGENTISTS

Emergentist approaches to the mind-body problem have been less prominent in the second half of the century than they were in the first, but they have continued to attract some interest. In this section we exam-

12. Wilfrid Sellars, "Science, Sense Impressions, and Sensa: A Reply to Cornman," *Review of Metaphysics* 24 (March 1971): 408.

13. Ibid., p. 409.

14. And other predicates; sensa are shaped as well as colored.

15. It follows from this, interestingly, that Sellars rejects the causal closure of the physical, in the sense in which this is understood by Kim.

ine two such approaches: one by the Nobel laureate neuroscientist Roger Sperry, and the other by Karl Popper.

Roger Sperry

Beginning in the 1960s, the neuroscientist Roger Sperry found it necessary to modify the currently dominant physicalist model of psychology by assigning a causal role to consciousness. Over the subsequent decades, this move became the basis for a rather expansive philosophy whose aims (as stated in the subtitle of a 1983 volume) include "merging mind, brain, and human values."[16] This philosophy, based on a version of emergent evolution, promises to overcome the conflict of science and religion, resolve the free will problem and the fact-value dichotomy, and provide the basis for a "naturalistic global ethic" that can "get ethico-religious values from science in a prescriptive sense."[17]

Some of these broader aims of Sperry's project go beyond what can be discussed here in any detail. His proposal concerning free will is straightforwardly compatibilist: "The whole process is still controlled or determined, but primarily by emergent cognitive, subjective intentions of the conscious/unconscious mind." With regard to the role of values, he argues that "if conscious mental values not only arise from but also influence physical brain action, it then becomes possible to integrate subjective values with objective brain function and its physical consequences." So far, what this gives us is that *a subject's perception of value* can affect behavior; this does not, of course, address the "fact-value dichotomy" as philosophers have understood it. He does move in that direction, however, when he claims that "Macromentalist theory . . . provides a master plan based in emergent evolution, which, though not preconceived but gradually self-determined in its design, is nevertheless replete with intrinsic directives for determining values."[18] Expanding on this, he claims that "the highest good, ultimate meaning, moral right and wrong, and so on are deter-

16. Roger Sperry, *Science and Moral Priority: Merging Mind, Brain, and Human Values* (New York: Columbia University Press, 1983).

17. R. W. Sperry, "Psychology's Mentalist Paradigm and the Religion/Science Tension," *American Psychologist* 43 (August 1988): 607–13; the citation is from p. 610. For another statement of the general character of Sperry's philosophy, see "In Defense of Mentalism and Emergent Interaction," *Journal of Mind and Behavior* 12 (Spring 1991): 221–45; material cited is on pp. 235–36.

18. Sperry, "Psychology's Mentalist Paradigm," p. 610.

mined in terms of concordance with the master plan for existence set by evolving nature, including human nature with its sociocultural as well as genetic values"; the final ethical criterion is found in "the 'common good,' ultimately in the perspective of the long-range evolving quality of the biosphere as a whole."[19]

Much of this, clearly, represents Sperry's personal perspective and need not be evaluated here. What is of present interest, however, is the character of the emergence-doctrine he articulates. It's clear, on the one hand, that he rejects both substance dualism and libertarian free will; these are his main points of disagreement with John Eccles and Karl Popper, whose views in other respects he finds quite congenial.[20] Since he rejects libertarian freedom, it is unlikely that he would see any need for emergent$_2$ entities or processes. On the whole, his preferred view would seem to be that consciousness in its various forms is, in our terms, emergent$_{1b}$—that it is an emergent property that, in virtue of its emergent causal powers, exercises a "downward" causal influence on the neural processes on which it supervenes. Thus, he states, "The principle of control from above downward, referred to as 'downward causation,' . . . says that we and the universe are more than just swarms of 'hurrying' atoms, electrons, and protons, that the higher holistic properties and qualities of the world to which the brain responds, including all the macrosocial phenomena of modern civilization, are just as real and causal for science as are the atoms and molecules on which they depend."[21] He also says, "The full explanation requires that one also take into account new, previously nonexistent, emergent properties, including the mental, that interact causally at their own higher level and also exert causal control from above downward."[22] A

19. Ibid., p. 611.

20. See Sperry, *Science and Moral Priority*, pp. 77–103. Sperry seems unaware of the unity-of-consciousness argument, though at times he comes tantalizingly close to it. Thus, he cites Eccles as asserting that "a key component of [Eccles's] hypothesis is that unity of conscious experience is provided by the mind and not by the neural machinery," and states himself to be "in full accord": "I too had made precisely the same point in 1952, stating: 'In the scheme proposed here, it is contended that unity in subjective experience does not derive from any kind of parallel unity in the brain processes. Conscious unity is conceived rather as a functional or operational derivative,' and 'there need be little or nothing of a unitary nature about the physiological processes themselves' " (ibid., p. 85). It doesn't seem to occur to Sperry to ask what it is that *has* the unitary experiences.

21. Sperry, "Psychology's Mentalist Paradigm," p. 609.

22. Ibid.

great many other statements by Sperry confirm the impression that he thinks of the emergent properties as causally effective.

Difficulties arise, however, concerning the relationship between this "downward causation" and the microdeterminism that he recognizes for the neural substrate. Sperry is unwilling to affirm that the laws governing the fundamental components must be different in the presence of the emergent properties. But this, as we have seen, is essential if there is to be genuine, "downward" causal influence by the emergent property. He writes, for instance:

> The expectation that downward macrodetermination should thus effect re-configurations . . . in the neuron-to-neuron activity of subjective mental states—or in the micro components of any macro phenomenon—indicates a serious misunderstanding of what emergent interaction is. From the start I have stressed consistently that the higher-level phenomena in exerting downward control do *not disrupt* or *intervene* in the causal relations of the lower-level component activity. Instead they *supervene* in a way that leaves the micro interactions, per se, unaltered.[23]

This naturally prompts a question: if the "downward control" does *not* affect the micro interactions, how is it causally effective? Sperry conveys his answer to this by means of an example:

> A molecule within the rolling wheel, for example, though retaining its usual inter-molecular relations within the wheel, is at the same time, from the standpoint of an outside observer, being carried through particular patterns in space and time determined by the over-all properties of the wheel as a whole. There need be no "reconfiguring" of molecules relative to each other *within the wheel itself.* However, *relative to the rest of the world* the result is a major "reconfiguring" of the space-time trajectories of all components in the wheel's infrastructure.[24]

It is true, of course, that the wheel's component molecules move differently in relation to the rest of the world because they are part of a rolling

23. Sperry, "In Defense of Mentalism," p. 230. This 1991 article is particularly important for understanding Sperry's view, because in it he attempts to improve on his earlier attempts at explaining the position, which he admits "have not been overly successful" (p. 223).

24. Ibid.; emphasis in original.

wheel than they might otherwise—if, for example, they were part of a fragment of metal lying on the ground. But this fails to illuminate the situation as regards consciousness for two reasons: First, if consciousness is to make a difference it has to make that difference in the *internal functioning of the brain itself*; absent this, there is no intelligible way in which consciousness could "reconfigure" the person's relationship to the rest of the world. Second, the macroscopic movements of the wheel as a whole are themselves quite thoroughly explicable in the reductionist style, in accordance with "bottom-up" microdeterminism. Given these two differences, the supposed example fails to illuminate the situation it is called upon to clarify.

Sperry, however, sees the matter quite differently. In an extended comment on this example, he writes:

> Whereas the "wheel rolling downhill" may be only an analogy in respect to consciousness, it is a direct, simple, objective, physical *example* in respect to the general principle of macrodeterminism and emergent causation. It illustrates one way in which nonreductive emergent properties determine the interactions of an entity as a whole at its own level, and also exert supervenient downward control, determining the space-time trajectories of its components at all lower levels.
>
> The rolling wheel example shows, further, that these emergent interactions are accomplished without disrupting the chains of causation among the sub-entities at their own lower levels, nor in their upward determination of the emergent properties. In other words, there is no breach in the previously posited physical determinism within the lower-level interactions.[25]

The reader may well be puzzled by Sperry's claim that the ordinary laws governing the motions of molecules fail to explain such a phenomenon as a wheel's rolling downhill. His justification for this is found in his assertion that "the lower level laws fail to include the complex, but specific, spacing and timing of the parts. These space-time, configurational, form or pattern factors are predicated to be causative themselves."[26] This, however, is simply confused. The laws of molecular motion don't *mention* the particular configuration of molecules in a rolling wheel; they couldn't pos-

25. Ibid., pp. 224–25.
26. Ibid., p. 225.

sibly do this, since they are quite general in nature. But they do, of course, *take account* of such "space-time, configurational, form or pattern factors" in predicting molecular movements. And by thus taking account of the configurational factors, extremely accurate predictions can often be made concerning rolling wheels, orbiting spacecraft, and the like. All this occurs within the framework of deterministic, "bottom-up" reductionist physics, and so the assumed need in such cases for "emergent properties" exerting "supervenient downward control" is shown to be an illusion. Emergence just is not as easy to demonstrate as this example suggests, and Sperry has not justified his claim that "emergent causation of this kind is ubiquitous, almost universal."[27]

Another favorite example concerns the relationship between the physical processes in a television receiver and the program content that is received. The physical processes operate according to the customary laws of physics and electronics, but those laws are unable to explain the program content. Here Sperry himself notes a flaw in the example: "The television analogy breaks down if pushed too far, of course, in that the superimposed programs of television are linearly traceable to the recording studio, whereas the brain, by contrast, is largely a self-programming, self-energizing system."[28] Quite so—but then Sperry owes us an account of how the brain, operating according to the *unaltered, invariant micro-laws* which govern its component parts, is able to generate its own programming. Sperry, however, simply notes that "the programs passing through the television monitor lack the internal interaction and competition of those of the brain, as well as the *self-developing, originative properties* and an internal selector of the program to be attended."[29] It is, of course, just these self-developing, originative properties that need to be accounted for *without allowing for any modification of or deviation from the micro-laws.* Instead of addressing this, Sperry walks away from his example at the crucial point.

I conclude, therefore, that Sperry's doctrine of emergence suffers from an incoherence that verges on contradiction. What he wants, one might say, is emergence$_{1b}$, with its emergent causal powers and its robust notion of downward causation. But his insistence that the micro-laws remain unaltered means that at most he can have emergence$_{1a}$, ordinary causal emer-

27. Ibid.
28. Sperry, *Science and Moral Priority*, p. 95.
29. Ibid.; emphasis added.

gence, according to which in the end everything is explicable—in principle, if not always in practice—in bottom-up terms.[30] One might wonder, though, how Sperry himself could have failed to notice this fact. I am not certain of the answer to this question, but I will suggest a possible answer—one that may apply to a number of other theorists in addition to Sperry. I suspect that the confusion arises because of a misleading suggestion, noted in the previous section, that is inherent in the notions of "levels" and "downward causation." As we've seen, the "levels" involved are levels of organization and integration: a living cell consists of the same old atoms and molecules, but organized and interacting in ways that differ profoundly from the organization and interaction of the same atoms and molecules as found in a chemical laboratory. And to say that the higher levels causally affect the lower levels can only mean that, because of the specific sort of organization and integration involved, the *behavior of the fundamental elements* is different than it would be if those elements continued to be governed by the old, non-emergent laws.

Suppose, however, that someone is misled by the terminology of levels and of downward causation into thinking of the different levels as concrete and as capable of exerting, on their own, distinct kinds of causal influence—more or less along the lines of the multistoried building mentioned above. Then the idea that the higher levels can causally influence the lower ones *without modifying the laws that operate on the lower levels* begins to seem coherent. Evidence that Sperry may be thinking this way is found in his assertion that "there exists within the cranium a whole world of diverse causal forces" equated with the various levels, and "it comes down to the issue of who pushes whom around in the population of causal forces that occupy the cranium."[31] In order to exert these "diverse causal forces," the different levels must have concrete existence, and if that is so they may indeed be able to "push around" the microelements—thus competing with, rather than superseding, the causal forces generated by those elements according to the ordinary laws of physics. But of course this whole way of looking at the matter is an illusion derived from the picture, which fundamentally misrepresents what is going on. To repeat just once more: *the only concrete existents involved are the ultimate constituents*

30. Timothy O'Connor attempts to construct an interpretation of Sperry's views that renders them consistent along the lines of our emergence$_{1b}$. I suspect, however, that he is being excessively charitable; he attributes ideas to Sperry that have no clear basis in Sperry's text. See O'Connor, "Emergent Properties," pp. 101–2.

31. Sperry, *Science and Moral Priority*, p. 32.

and combinations thereof, the only causal influences are those of the ultimate constituents in their interactions with each other, and the only way the "higher levels" can make a difference is by altering or superseding the laws according to which they interact.

But Sperry fails to see this, resulting in the incoherence we have noted. Nevertheless, his protest against the received physicalist viewpoint in psychology retains considerable force in spite of his failure to articulate his alternative in a way that is philosophically coherent. And the favorable reception of his perspective by a segment of the psychological community suggests that this community is to a significant degree open to antireductionist viewpoints. It should further be noted that Sperry's argument for the causal efficacy of consciousness is by no means tied to his particular conclusions concerning free will, values, and religion; those who disagree with those conclusions can still rightly perceive him as an ally in opposing physicalism and reductionism.

Karl Popper

Probably the most significant philosophical voice for emergentism in the last half-century has been that of Karl Popper. Interestingly, Popper is discussed more extensively by Sperry than any other philosopher,[32] and Popper in turn indicates awareness of Sperry's work.[33] Popper's views on this topic have been less influential than his well-known contributions to the study of scientific method, but they provide some provocative insights and are well worth discussing.

Popper refers to the emergent evolutionists, and indicates general agreement with their approach. He opposes the philosophical dogma expressed by the saying There is no new thing under the sun, and suggests instead "that the universe, or its evolution, is creative, and that the evolution of sentient animals with conscious experiences has brought about something new" (p. 15). He never gives a precise analysis of "emergence" (Popper frequently expresses his distaste for discussing the meanings of words), but by attending to context it is possible to get a fair idea of what he has in mind. In general, one may say that for Popper the point of emer-

32. In *Science and Moral Priority*, chap. 6.

33. See Karl R. Popper and John C. Eccles, *The Self and Its Brain: An Argument for Interactionism* (New York: Springer-Verlag, 1977), p. 209. Part I of this book, written by Popper, is the primary source for his views on emergentism and the mind-body problem. (Page references in this section are to this work.)

gence is that the course of the universe's development gives rise to new characteristics, not able to be anticipated or predicted on the basis of what has gone before. He lays considerable stress on *indeterminism* in this connection, apparently because this guarantees the unpredictability, and therefore the genuine novelty, of the outcome of the process. (This contrasts with Sperry, who, as we have seen, emphasizes that his concept of emergence is fully compatible with determinism.) He considers the objection that "if something new seems to emerge in the course of the evolution of the universe . . . then the physical particles or structures involved must have possessed beforehand what we may call the 'disposition' or 'possibility' or 'potentiality' or 'capacity' for producing the new properties, under appropriate conditions" (p. 23). His reply is that this is either trivial or misleading: trivial, in the sense that there obviously must be something in evolutionary history that preceded and prepared the way for the appearance of the new feature; but misleading if it is assumed that the precursor must be something very similar to the new, emergent characteristic. (It is for this reason that he opposes panpsychism, which he treats as a rival and alternative to emergentism.) He writes, "We know of processes in nature which are 'emergent' in the sense that they lead, not gradually but by something like a leap, to a property which was not there before" (p. 69).

Popper does appeal to the notion of "downward causation," and it has to be said that his employment of this idea is not as careful or as precise as one might wish. He wants to establish the idea for use in connection with mind-body interaction, but he applies it very broadly, sometimes in contexts where the implications are far from clear. Thus he says,

> There are many . . . macro structures which are examples of downward causation: every simple arrangement of negative feedback, such as a steam engine governor, is a macroscopic structure that regulates lower level events, such as the flow of the molecules that constitute the steam.
>
> Downward causation is of course important in all tools and machines which are designed for some purpose When we use a wedge, for example, we do not arrange for the action of its elementary particles, but we use a structure, relying on it to guide the actions of its constituent elementary particles to act, in concert, so as to achieve the desired result. (p. 19)

It is a little difficult to see what is meant here by "downward causation": presumably Popper does not want to claim that the higher-level structures in these cases operate in some way other than through molecular interactions! In general, downward causation as Popper speaks of it does not necessarily imply "novel causal influence" on the part of the emergent feature, though neither does it exclude this. Nor does downward causation exclude in principle the possibility of explanation in terms of "bottom-up" microcausation (see p. 23), though he clearly does not think such explanations can be given in all cases.

How then does Popper's theory of emergence stand in relation to the alternative conceptions developed in the first section of this chapter? Probably the best answer is that he is thinking in terms of causal emergence, Searle's emergence$_1$. Popper's use of "emergence" does not necessarily imply emergence$_{1b}$, the emergence of novel causal powers, although some of the cases he mentions, especially those involving consciousness, clearly are in that category. And his advocacy of libertarian freedom means that the mind, in particular, has features that are emergent$_2$.

The mind, of course, emerges from the functioning of the brain and is closely tied to it; Popper conjectures that "the flawless transplantation of a brain, were it possible, would amount to a transference of the mind, of the self" (p. 117). He is skeptical about the existence of minds after death (see p. 556), though he never claims this is logically impossible. He thinks "the very idea of substance is based on a mistake" (p. 105), and does not use the word "substance" in stating his own view.[34] Nevertheless, it seems clear that he thinks of the mind or self as a continuing individual entity, distinct from the brain and interacting with it; in our terms, he is a substance dualist. While affirming the liaison between the self and its brain to be extremely close, he nevertheless warns against assuming "too close and too mechanical a relationship" (p. 118). As evidence against such an assumption, he points out that "at least some outstanding brain scientists have pointed out that the development of a new speech centre in the undamaged hemisphere reminds them of the reprogramming of a computer. The analogy between brain and computer may be admitted; and it may be pointed out that the computer is helpless without the programmer" (p. 119). In some cases brain functions stand in a one-to-one relationship with experience, but in other cases "this kind of relationship cannot be empiri-

34. Popper writes: "I conjecture that the acceptance of an '*interaction*' of mental and physical states offers the only satisfactory solution of Descartes's problem" ("Of Clouds and Clocks," in *Objective Knowledge: An Evolutionary Approach* [Oxford: Clarendon, 1972], p. 252); he does not say what the "mental states" are states *of.*

cally supported." In another passage he returns to the computer metaphor, and also displays some of his affinities with traditional dualism: "I have called this section 'The Self and Its Brain,' because I intend here to suggest that the brain is owned by the self, rather than the other way round.[35] The self is almost always active. The activity of selves is, I suggest, the only genuine activity we know. The active, psycho-physical self is the active programmer to the brain (which is the computer), it is the executant whose instrument is the brain. The mind is, as Plato said, the pilot" (p. 120). Still another passage puts the emergence of mind in a broad evolutionary perspective:

> The emergence of full consciousness, capable of self-reflection, which seems to be linked to the human brain and to the descriptive function of language, is indeed one of the greatest of miracles. But if we look at the long evolution of individuation and of individuality, at the evolution of a central nervous system, and at the uniqueness of individuals . . . then the fact that consciousness and intelligence and unity are linked to the biological individual organism (rather than, say, to the germ plasm) does not seem so surprising. For it is in the individual organism that the germ plasm—the genome, the programme for life—has to stand up to tests (p. 129).

EMERGENT DUALISM

In the remainder of this chapter I will sketch out a theory of the mind which makes the mind both emergent$_{1b}$, since it is endowed with novel causal powers, and also emergent$_2$, since it possesses libertarian free will. I shall not claim either that this theory provides the only possible solution to the problem of the nature of persons, or that it is without difficulties of its own. I will count myself successful if I can leave the reader with the perception that this is a view that merits further consideration—that it may offer a way forward through the thicket of difficulties which perplex us.

We begin by stipulating that we take the well-confirmed results of natural science, including research on neurophysiology, just as we find them. Attempts to resolve the problem through a nonrealistic interpretation of the sciences, as in idealism and some forms of phenomenology, are deeply

---

35. It was at Eccles's suggestion (see p. 473) that this was made the title for the book as a whole.

implausible and provide no lasting solution to our problems.[36] We need not assume that the sciences give us a *complete account* of the nature of the world, even an "in-principle" complete account. But what they do give us, in the form of their well-confirmed results, must be acknowledged as in the main true (or at least approximately true), and as informative about the real nature of things.

But if our theory should be realist about the results of the sciences, it should also be realist about the phenomena of the mind itself. John Searle has noted that a great deal of recent philosophy of mind is extremely implausible because of its denial of apparently "obvious facts about the mental, such as that we all really do have subjective conscious mental states and that these are not eliminable in favor of anything else."[37] It's true that we do not, in the case of the mind, have well-confirmed scientific theories comparable to the powerful theories that have been developed in the physical sciences. But we do have a vast amount of *data* concerning mental processes, events, and properties, and we should begin with the presumption that we are going to take these data as they stand (subject of course to correction), rather than truncate them in order to tailor them to the requirements of this or that philosophical scheme.

So far, perhaps, so good. But stating that we are realists both about the physical and about the mental brings to the fore once again the vast differences between the two: the chasm opens beneath our feet. Cartesian dualism simply accepts the chasm, postulating the soul as an entity of a completely different nature than the physical, an entity with no essential or internal relationship to the body, which must be added to the body *ab extra* by a special divine act of creation. This scheme is not entirely without plausibility, at least from a theistic point of view, but I believe (and have argued above) that it carries with it serious difficulties.

In rejecting such dualisms, we implicitly affirm that *the human mind is produced by the human brain and is not a separate element "added to" the brain from outside.* This leads to the further conclusion that mental properties are "emergent" in the following sense: they are properties that manifest themselves when the appropriate material constituents are placed in

36. The mind-body problem arises, in large part, because of the apparent incongruity between the well-confirmed results of the natural sciences and what seems experientially to be the case with regard to the mind. Giving a nonrealistic interpretation of the sciences simply moves the incongruity to another place, between the manifest content of the scientific disciplines and the philosophical interpretation that is given of that content.

37. Searle, *Rediscovery of the Mind*, p. 3.

special, highly complex relationships, but these properties are not observable in simpler configurations nor are they derivable from the laws which describe the properties of matter as it behaves in these simpler configurations. Which is to say: *mental properties are emergent$_{1b}$*; they involve emergent causal powers that are not in evidence in the absence of consciousness.

But while property emergence is necessary for the kind of view being developed here, it is not sufficient. For the unity-of-consciousness argument spelled out in Chapter 5 claims to show not only that the properties of the mind cannot be explained in terms of the properties exhibited by matter in simpler, nonbiological configurations; it claims that these properties cannot be explained in terms of—that is, they are not logical consequences of—*any* combination of properties of, and relations between, the material constituents of the brain. A conscious experience simply *is* a unity, and to decompose it into a collection of separate parts is to falsify it. So it is not enough to say that there are emergent properties here; what is needed is an *emergent individual*, a new individual entity which comes into existence as a result of a certain functional configuration of the material constituents of the brain and nervous system. Endowed, as we take it to be, with libertarian freedom, this individual is able, in Searle's words, to "cause things that could not be explained by the causal behavior of the neurons"; in virtue of this, consciousness is indeed emergent$_2$. As an analogy which may assist us in grasping this notion, I suggest the various "fields" with which we are familiar in physical science—the magnetic field, the gravitational field, and so on. A magnetic field, for example, is a real, existing, concrete entity, distinct from the magnet which produces it. (This is shown by the fact that the field normally occupies—and is detectable in—a region of space considerably larger than that occupied by the magnet.) The field is "generated" by the magnet in virtue of the fact that the magnet's material constituents are arranged in a certain way— namely, when a sufficient number of the iron molecules are aligned so that their "micro-fields" reinforce each other and produce a detectable overall field. But once generated, the field exerts a causality of its own, on the magnet itself as well as on other objects in the vicinity. (In an electric motor, the armature moves partly because of the magnetic fields produced by itself.) Keeping all this in mind, we can say that *as a magnet generates its magnetic field, so the brain generates its field of consciousness.* The mind, like the magnetic field, comes into existence when the constituents of its "material base" are arranged in a suitable way—in this case, in the extremely complex arrangement found in the nervous systems of humans and other animals. And like the magnetic field, it exerts a causality of its own; certainly on the brain itself, and conceivably also on other minds (telepathy) or on other aspects of the material world (telekinesis).

To be sure, this analogy has its limitations, and it is important to keep in mind that it *is* an analogy rather than an attempt at causal explanation. In order to take the analogy in the right way, try to think of the generation of the magnetic field naively, somewhat as follows: to begin with, we have a coil of wire, with no associated magnetic field. We then cause an electric current to pass through the wire, and the presence of the current "causes a field of force to appear in previously empty space."[38] A new individual (however short-lived) has come into existence. In a somewhat similar way, on the present hypothesis, the organization and functioning of the nervous system bring into existence the "field of consciousness." One difference, of course, is that we know quite accurately the necessary and sufficient conditions for the generation of a magnetic field, whereas we know very little about the conditions for the emergence of consciousness.

Admittedly, the suggested way of thinking about the magnetic field *is* naive, and one might reasonably ask what becomes of the analogy when we view it in terms of the *real* ontology of fields, as opposed to the oversimplified version just presented. The problem with this is that the "real ontology of fields" has been a matter of debate ever since the introduction of the field concept (and its conceptual cousin, action at a distance) into physics.[39] As for the situation today, we need only recall the continuing controversy about the right way to interpret quantum mechanics. As Richard Feynman has observed, quantum electrodynamics is a theory that makes remarkably precise predictions (the theoretically calculated value for Dirac's number agrees with experiment to ten decimal places), but "nobody understands . . . why Nature behaves in this peculiar way."[40] If this situation is someday resolved, so that a stable consensus on the ontology of fields is arrived at, it will certainly be necessary to reexamine the use of the field analogy in the philosophy of mind in the light of our new knowledge.

The limitations of the analogy are further shown in the fact that the

38. See Mary Hesse, *Forces and Fields: The Concept of Action at a Distance in the History of Physics* (London: Thomas Nelson, 1961), p. 250.

39. The history is well documented in Hesse, *Forces and Fields*.

40. Richard Feynman, *QED: The Strange Theory of Light and Matter* (Princeton: Princeton University Press, 1985), p. 10.

properties of the magnetic field and the other fields identified by physics do not seem to be emergent in the strong sense required for the properties of the mind. Nor does it seem that these fields possess the kind of unity that is required for the mind, as shown by the unity-of-consciousness argument. And there is no reason to suppose that the fields of physics are endowed with inherent teleology, much less libertarian freedom. The analogy with the magnetic field is of some value in helping us to conceive of the ontological status of the mind according to the present theory. But it is *only* an analogy, and as such it can't bear the full weight of the theory, which must rather commend itself in virtue of its inherent advantages over both materialism and Cartesian dualism.

The theory's advantages over Cartesian dualism result from the close natural connection it postulates between mind and brain, as contrasted with the disparity between mind and matter postulated by Cartesianism. In view of this close connection, it is natural to conclude that the emergent consciousness is itself a spatial entity. If so, it would seem that emergent dualism is well placed in relation to Kim's "pairing problem." That problem, it will be recalled, asks about the basis of the connection whereby a particular soul and body are able to causally interact with each other. Timothy O'Connor speculates in this regard that "what is needed, perhaps, is an asymmetrical dependency-of-existence relation—most likely, this body (at the right stage of development) generating that mind. If this kind of baseline dependency relation is intelligible, the fact that these two entities should also interact in more specific ways over time does not seem to be a *further* mystery."[41] This seems correct, and it allows us to say a bit more about the spatial nature of the emergent mind: the volume of space within which the emergent mind exists must be *at least* sufficient to encompass those parts of the brain with which the mind interacts. It was argued in the previous chapter that the difficult of conceiving mind-body interaction is not a conclusive objection against Cartesian dualism. But there seems little doubt that such interaction is *more readily* intelligible for emergent dualism, and this would seem to constitute a significant advantage for the latter theory.

There is evidence both from subhuman animals and from human be-

41. Timothy O'Connor, "Comments on Jaegwon Kim, 'Causality and Dualism' " (paper delivered at the University of Notre Dame, March 7, 1998). In fairness, I should quote also O'Connor's next remark: "But the idea of a natural emergence of a whole substance is perhaps a lot to accept."

ings (e.g., commissurotomy) that the field of consciousness is capable of being divided as a result of damage to the brain and nervous system.[42] This fact is a major embarrassment to Cartesianism, but it is a natural consequence of emergent dualism. Beyond this, the theory makes intelligible, as Cartesian dualism does not, the intimate dependence of consciousness and mental processes on brain function. The detailed ways in which various mental processes depend on the brain must of course be discovered (and are in fact being discovered) by empirical research. Philosophy should be wary of attempting to anticipate these conclusions, lest it reenact the tragicomedy of the pineal gland. But there is no reason to think the kind of theory here proposed will have any difficulty in accommodating the results as they emerge. It needs to be kept in mind, however, that the mind is not merely the passive, epiphenomenal resultant of brain activity; instead, the mind actively influences the brain at the same time as it is being influenced by it. And, finally, this theory is completely free of embarrassment over the souls of animals. Animals have souls, just as we do: their souls are less complex and sophisticated than ours, because generated by less complex nervous systems.

The theory's advantages over materialism will depend on which variety of materialism is in view. As compared with eliminativist and strongly reductive varieties of materialism, our theory has the advantage that it takes the phenomena of mental life at face value instead of denying them or mutilating them to fit into a Procrustean bed. In contrast with mind-body identity theories and supervenience theories that maintain the "causal closure of the physical," the view here presented recognizes the necessity of recognizing both teleology and intentionality as basic-level phenomena; they are not the result of an "interpretation" (by what or by whom, one might ask?) of processes which in their intrinsic nature are

---

42. There are, of course, numerous organisms (e.g., starfish) that can be divided into parts, with each part subsequently developing into a complete organism. As yet (and many of us hope this will not change) nothing of the sort has been done with human beings, though recent examples of the cloning of mammals suggest that the cloning of humans is a technologically possible. Even more telling are the split-brain data. Eccles admits that in split-brain cases "there is remarkable evidence in favour of a limited self-consciousness of the right hemisphere" (*Evolution of the Brain: Creation of the Self* [London: Routledge, 1989], p. 210). This is especially significant coming from Eccles, who is essentially a Cartesian dualist: it is hardly intelligible that a Cartesian consciousness should be divided by an operation on the brain, so Eccles's admission has to reflect strong empirical pressure from the experimental data.

neither purpose-driven nor intentional. The view proposed here has more affinity with "property dualism" and views which postulate a strong form of property emergence—but these already are views to which many will hesitate to accord the label "materialist." Be that as it may, the present view differs from property dualism and property-emergence views in its postulation of the mind as an *emergent individual*, thus providing it with an answer, which those views lack, to the problem posed by the unity-of-consciousness argument.

The resemblance to property-emergence views does, however, suggest a suitable name for the theory. At one time I referred to it simply as "emergentism,"[43] but that label could lead to misunderstanding because it is most commonly used for theories of property emergence. I suggest, then, "emergent dualism" as a name which brings to the fore both the theme of emergence and the undeniable affinities between the "soul-field" postulated here and the mind as conceived by traditional dualism.

I have described the advantages of emergent dualism, but what of the costs? So far as I can tell, there is only one major cost involved in the theory, but some will find that cost to be pretty steep. The theory requires us to maintain, along with the materialists, that the potentiality for conscious life and experience really does exist in the nature of matter itself.[44] And at the same time we have to admit, as Colin McGinn has pointed out, that we have no insight whatever into how this is the case.[45] It is not necessary to endorse McGinn's assertion that the brain-mind link is "cognitively closed" to us—that is, that human beings are inherently, constitutionally incapable of grasping the way in which matter produces consciousness—though that possibility deserves serious consideration. And yet, in purely physiological terms, what is required for consciousness—or at least, some kind of sentience—to exist, must not be all that complex,

43. See Hasker, "The Souls of Beasts and Men," *Religious Studies* 10 (1974): 265–77, and "Emergentism," *Religious Studies* 18 (1982): 473–88.

44. David Chalmers has suggested in discussion that the emergence doctrine would not force us to revise our conception of matter, if we consider the laws of emergence as contingent laws of nature that merely specify what happens under given circumstances. This seems to be correct, but I find this conception of laws implausible; it immediately invites the question, What is the ontological grounding of the causal powers involved? The view taken here is that laws of nature formulate causal powers that inhere in the natures of natural causal agents.

45. See Colin McGinn, *The Problem of Consciousness* (Oxford: Blackwell, 1991), chap. 1.

since the requirements are apparently satisfied in relatively simple forms of life. As McGinn puts it, "In the manual that God consulted when he made the earth and all the beasts that dwell thereon the chapter about how to engineer consciousness from matter occurs fairly early on, well before the really difficult later chapters on mammalian reproduction and speech."[46]

While emergent dualism shares with (nonreductive) materialism the claim that ordinary matter contains within itself the potentiality for consciousness, it actually goes some way beyond materialism in the powers it attributes to matter. For standard materialism, the closure of the physical guarantees that consciousness does not "make a difference" to the way matter itself operates; all of the brain-processes are given a mechanistic explanation which would be just the same whether or not the processes were accompanied by conscious experience. Emergent dualism, on the other hand recognizes that a great many mental processes are *irreducibly* teleological, and cannot be explained by or supervenient upon brain processes that have a complete mechanistic explanation. So the power attributed to matter by emergent dualism amounts to this: when suitably configured, it generates a field of consciousness that is able to function teleologically and to exercise libertarian free will, and *the field of consciousness in turn modifies and directs the functioning of the physical brain*. At this point, it must be admitted, the tension between the apparently mechanistic character of the physical basis of mind and the irreducibly teleological nature of the mind itself becomes pretty severe, and the siren song of Cartesian dualism once again echoes in our ears.

Brian Leftow has suggested a reading of Aquinas which in a certain way bridges the difference between the emergent dualism advocated here and creationism. He writes:

> By a continuous rearranging of live matter (we'd now say: by the brain's development), the human fetus becomes able to host the human soul, i.e. develops the full material base for the capacity to think in (what Thomas thinks is) the soul-requiring way. At that point, the capacity becomes present, and with it the individual(s) it requires. This happens in so lawlike a way as to count as a form of natural supervenience. So if we leave God out of the picture, the Thomist soul is an "emergent individual." . . . The law-

46. McGinn, *Problem of Consciousness*, p. 19. (McGinn's fairly frequent references to God are to be taken heuristically and not as expressions of actual belief.)

like way brain-development leads to souls' appearance may make it look like the brain's development causally accounts for the soul's appearance, but in this one case, Thomas is (as it were) an occasionalist.[47]

Here I am reminded of a choice remark of Fred Freddoso: "I love occasionalism, but I'd hate to have to believe it!" Just as, in general, there are powerful reasons for preferring a generally realist account of creaturely causation to occasionalism, so also in this specific case. To be sure, if we had compelling reasons to suppose that the emergence of the soul through natural causation is impossible (as opposed to merely being difficult for us to understand), the occasionalist option would deserve a serious second look. But lacking such reasons, causal realism about the emergence of the soul is the better way to go.[48]

Leftow, however, presents the following dilemma:[49] On the one hand, if the "field of consciousness" consists of previously existing "things or stuff," the field is material after all. If it does not, then the emergence of the field amounts to creation *ex nihilo*, and Leftow thinks theists may hesitate to suppose that God will or can share his power to create with the brains from which the field of consciousness emerges. My response is to reject both alternatives. If the field consisted of previously existing stuff—quarks, for instance—it would be vulnerable to the unity-of-consciousness argument previously utilized against materialism. And on the other hand, the power to create *ex nihilo* is indeed a traditional attribute of God, something I am unwilling to attribute to a mere creature. What I want to say is simply that the generation of the field of consciousness is a consequence of the natural powers with which the stuff of life is endowed. And I deny that when a created thing produces another created thing through the exercise of its natural powers, the thing that does the producing should be said to have created something *ex nihilo*. But the *way* in which the production of the field occurs is something it is best for now to leave undecided. (A bit more will be said about this in the last section of this chapter.)

It should be clear by this time that emergent dualism has no special

47. Leftow, comment delivered at the University of Notre Dame, March 5, 1998.
48. There is a puzzle for this kind of occasionalism in the phenomenon of the *perfectly symmetrical fission* of an ensouled organism: which twin gets the original soul, and which the reproduction? It doesn't look as though this could be settled by any universal law of supervenience; and if not, the occasionalist model breaks down.
49. Leftow, comment.

problem with biological evolution. Previously it was argued that hard-line Darwinian evolution, constrained by the closure of the physical, cannot possibly be correct; it fails entirely to account for any correspondence between physical reality and the content of our subjective experience. Emergent dualism does not have this problem: since conscious states are causally effective, they are also subject to Darwinian selection. And since at least some conscious states are inherently teleological, it is unnecessary to follow the ingenious but implausible attempts of Darwinists to account for all apparent teleology by selection explained in purely mechanistic terms. Given that the conscious states are teleological, Darwinian processes can indeed operate to select those which are more effective in reaching goals conducive to survival and reproduction.[50] Whether this liberalized Darwinism is sufficient to account for all the phenomena of life (including human life) as we know it is a fascinating question, one that needs to be investigated rather than dogmatized about; intriguing problems, both philosophical and empirical, abound in the neighborhood.

FURTHER PROBLEMS

We conclude this chapter with four intriguing questions suggested by the foregoing considerations. None of the questions will receive a definitive answer here; the last will become the topic of the final chapter.

Mind-Brain Correlation?

Just how close should we suppose the correlation of mental states and brain-states to be? Traditional dualists have usually asserted that certain mental processes—typically the higher reasoning processes—are lacking any neural correlates. To say that this is implausible, in the light of current neurological knowledge, is an understatement. PET scans, for instance, give very strong evidence that intellectual activities of all kinds involve neural processing. But how tight is the connection? We've noted Popper's warning against assuming "too close and too mechanical a relationship" between mind and brain, as well as his observation that in some cases the assumption of one-to-one correlation between mind-states and brain-states "cannot be empirically supported." Empirical evidence is highly rel-

---

50. This seems to be roughly Popper's view of evolution.

evant here, but it's likely that for the foreseeable future such evidence won't provide a definitive answer. Given this uncertainty, even nonmaterialists are motivated to make the connection as tight as possible, so as to avoid giving offense unnecessarily to the materialist mainstream.

Consider, then, what I am going to term a *strict correlation* between mental states and brain-states. This will not be literally a one-to-one correlation; it is generally agreed that neural states are more "fine-grained" than conscious states, and minor variations on the brain level may make no difference at the level of mind. So let us define "strict correlation" as follows:

> Mental states and brain-states are strictly correlated iff for every mental state there is simultaneously a corresponding brain-state, and for every brain-state of the relevant sort there is simultaneously a corresponding mental state, and any significant alteration in either the brain-state or the mental state is matched by a corresponding variation in its correlate.

This is still fairly rough—both "relevant sort" and "significant alteration" call for further clarification—but perhaps it's enough to get on with. Given the general approach set out in the preceding section, is strict correlation plausible? In order to facilitate the discussion, I offer the following diagram:

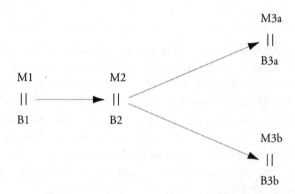

Here 1, 2, and 3 represent moments of time separated by brief intervals; M1, B1, and so on represent the mental states and brain-states of a person at those moments; and the double lines represent the relation of strict correlation, as defined above. Moment 2 is the time at which a free choice is made, leading either to the subsequent state M3a/B3a or to M3b/B3b. So now we ask: Is this schema coherent? And if coherent, is it plausible?

An initial observation is that, in order for strict correlation to obtain, the interaction between mental states and brain-states must be instantaneous. Of course any change, whether in the brain or in the mind, will take some time. But there can't be any time gap between a mental event and the correlate brain-event (or vice versa); otherwise we would have, for however brief a time, an alteration in the mental state with no alteration in the corresponding brain-state. Now the idea of instantaneous interaction may strike us as problematic, yet perhaps we should not reject it out of hand. (Something very similar to this is, after all, assumed in supervenience theory.) So far, then, the hypothesis of strict correlation seems viable.

There is an argument, based on our diagram, for the conclusion that strict correlation is not coherent. Recall that 2 is the moment at which a *free choice* is made. This choice, furthermore, is assumed to originate in the conscious mind; it is the mind's decision that determines whether the person ends up at M3a/B3a or M3b/B3b. But if this is so, then the state of the mind at 2 must be *different* if the "a decision" is made than if the "b decision" is made; in either case the mind has begun to "steer" the organism in one direction or the other. But it can't be the case that *both* of these mental states (call them "M2a" and "M2b") are the appropriate correlates for the same brain-state, namely B2. So strict correlation does not obtain.

The argument fails, of course, and the reason it fails is that the moment 2 hasn't been specified clearly enough. Is 2 the last instant *before* the "parting of the ways" of the paths leading to M3a/B3a and M3b/B3b? In that case, the mental state at 2 is indifferent as between the alternatives, and there is only one M2, not M2a and M2b. But if 2 is an instant *after* the two paths diverge, then there are indeed two mental states, M2a and M2b, but in that case there will also be two corresponding brain-states, B2a and B2b. So the coherence of strict correlation is maintained.

Upon reflection, however, I must confess that the notion of instantaneous interaction between distinct substances strikes me as implausible. Suppose an impulse of some sort arises in the mind: isn't it plausible that it takes some time, however brief, to overcome the "inertia" of the previous brain processes so that the brain responds appropriately? Or suppose the change comes in the opposite direction: a stimulus originating in the brain breaks in upon a train of thought. Isn't it plausible that in this case it takes some short interval to overcome what may be termed our "mental inertia," before the mind responds to the stimulus? In either case, one might suppose that there is first a very slight change in one of the correlates matched instantaneously by a slight change in the other, and then a larger, and more perceptible, alteration in each. But this seems to be an ad hoc expedient for saving what was, after all, no more than a plausible conjecture to begin with. In the end, the notion of strict instantaneous correlation strikes me as doubtfully consistent with the claim that we have here two distinct substances, each capable of originating changes and acting upon the other.

What is the Nature of Mind-Brain Interaction?

The issues raised in the previous section lead us naturally on to investigate more closely the nature of mind-brain interaction, according to emergent dualism. In general, there are four types of causal processes to be considered: brain-brain, brain-mind, mind-mind, and mind-brain. What can we say about the different types?

One question of undoubted interest is whether the various types of causation are deterministic or indeterministic. The brain-brain causal processes should no doubt be seen as following the ordinary laws of physics, and as deterministic except for the element of indeterminism implied by quantum mechanics. Brain-mind processes involve both the generation of the mind by the brain and the specific modification of mind-states by information from the brain. In the first instance, the generation of the field of consciousness should no doubt be thought of as deterministic; given a particular functional brain configuration, a particular sort of consciousness is bound to result. Given that the field of consciousness is "in place," the effects of brain on mind should also be viewed deterministically, but with this qualification: the resulting conscious state depends not only on the brain-state but on the mind's own internal evolution, so that one can't say in general that a given brain input inevitably produces a certain mental effect. Mind-mind causal processes are, plausibly, the place to find the causal indeterminism which is essential to free will and also to rationality. And finally, mind-brain causal influence should be thought of as deterministic, with the qualification that the resulting brain-state depends both on the internal evolution of the brain-states and on the influence thereon of the conscious mind.

All this, of course, leaves the precise nature of the causal processes involving the conscious mind very much a mystery. That this should be so is unavoidable in the present state of our knowledge. But I will hazard one further conjecture: on the hypothesis of emergent dualism, it is almost inevitable to conceive of these interactions as involving *exchange of energy* between mind and brain. It is predictable that this will provoke the objection that the mind, the field of consciousness, is "physical after all." My response to this is that little hangs on a name: if philosophers are prepared to stretch the meaning of "physical" to encompass everything that has been said here about the field of consciousness, then so be it. What is *not* acceptable, however, is for someone to take the claim, thus arrived at, that "the mind is physical" and use it as a premise from which to infer characteristics of the conscious mind that are contrary to the ones postulated in this chapter. The distinction between mind and body as conceived of here is simply different from that contemplated either by Cartesian dualism or by contemporary forms of materialism; the new wine can't safely be kept in those old, dry wineskins.

Is Life an Emergent?

Among the points on which emergent evolutionists differ among themselves is the question about the *number of levels* at which emergence occurs. Some have a small number of levels, while for others emergence is pervasive and the number of levels can be multiplied almost indefinitely. Of course, this question has a definite meaning only if the kind of emergence is specified: the weaker the notion of emergence, the easier it is to multiply levels. Suppose, then, we focus on the notion of emergence$_{1b}$, where the emergent feature exerts novel causal influence and modifies the behavior of the underlying elements. Clearly emergent dualism is committed to the claim that consciousness is emergent$_{1b}$—but is this the *only* example? In particular, what about biological life? Is life, also, an emergent feature?

There seem to be three main possibilities here. The first is that biological life, as such, is susceptible to complete mechanistic explanation in physico-chemical terms, and it is only with the advent of consciousness (in however primitive a form) that we have the emergent causal powers that characterize emergence$_{1b}$. This hypothesis has the advantage of clashing as little as possible with the mechanistic assumptions of most contemporary biology. (It also is congenial to Cartesian dualism, which may or may not speak in its favor.) The disadvantages, as I see them, are two: First, it intensifies the apparent incongruity between the consciousness which emerges and that from which it comes. Second, it is a fact that many living systems give us the overwhelming impression of functioning in a way that is purpose driven, but this hypothesis either disregards this impression or is forced to classify it as an illusion. Darwinists, to be sure, have their explanation of the illusion in terms of the "blind watchmaker"—but if we were to concede all of their claims (many of which are quite unsubstantiated by evidence) we should have abandoned this enterprise long ago. Simply put, even very elementary living things seem more like us, and less like pieces of clockwork, than this hypothesis seems to allow.

A second possibility is that life is indeed emergent$_{1b}$ and that the emergence of life *coincides with* the emergence of consciousness, in however elementary or primitive a form, so that *all* living things, no matter how humble, have within themselves some vestige of awareness. This rather bold conjecture has the great advantage that it is able to take at face value what seems to be the inherent teleology of living creatures, a feature that neither the "blind watchmaker" of Darwinism nor even the watchmaker God of some traditional theism seem to do justice to. On the other hand, the cost for this is the attribution of mentality to creatures which, even on the most generous estimate, give us no visible indication of possessing it. And by doing this the hypothesis moves us quite some distance in the direction of panpsychism—which may or may not be a disadvantage, but panpsychism is unquestionably a view to which many people would prefer not to be committed.

If neither of these views is especially appealing, there is a third alternative: Life involves emergent causal powers, and as such is emergent$_{1b}$; furthermore, the emergence of life is a distinct level of emergence which does not, as such, involve consciousness. This has the advantage of the previous view that it allows us to take seriously the appearance of inherent teleology in living things. It also has the advantage of enabling us to take a commonsense approach to the attribution of consciousness: where there is some evidence of consciousness (and of course opinions vary considerably as to what evidence is relevant), we can ascribe it, and where there is not we may withhold the attribution. The cost in this case lies in the need to spell out, in at least some minimal fashion, what is involved in the emergence of life and why a mechanistic explanation will not suffice.

These, then, seem to be the alternatives. Each has something in its favor, and none is without disadvantages. Each is consistent with the hypothesis of emergent dualism, and (so far as I can tell) also with the available empirical evidence. Beyond this point, I see at present no promising avenue by which to proceed.

Is Survival Possible?

Finally there is a question of considerable interest to emergent dualists as well as to others: What are the possibilities for survival of biological death? The belief in such survival and/or the desire to provide for its possibility have always been strongly associated with mind-body dualism, though there have also been dualists who have not shared such motivation. On the face of it, emergent dualism seems less well placed than either Cartesian or Thomistic dualism to return a favorable answer: a soul which emerges from the functioning of a biological organism, and is itself unable to function without support from the organism, seems a dubious candidate for immortality. On the other hand, materialistic views seem even less promising in this regard, yet there have been a number of attempts to make room for postmortem survival consistent with materialism. No doubt definitive philosophical answers to these questions will continue to elude us, but they will be made the topic of intensive discussion in the final chapter.

CHAPTER EIGHT

Prospects for Survival

It matters to us what happens to us after we die. Concern about this has both motivated religious belief and been motivated by it, but it is by no means limited to the religious. A few years ago an advertisement offered the opportunity, for a nominal fee, to have an ounce of one's ashes "placed aboard a rocket and sent to the sun where they will be transformed into pure energy, or sunshine, and radiated throughout the universe at the speed of light."[1] A more expansive hope is offered by cryonics, a practice in which persons immediately upon their decease are frozen at very low temperatures in the expectation that future science will learn how to revive them, heal their fatal illnesses, and enable them to resume their lives.[2] Science fiction, for the most part relentlessly secular in its assumptions, nevertheless demonstrates a fascination with the postmortem survival of human beings, usually by cybernetic means. In John Varley's novel *Wizard*,[3] the mental lives of select individuals are continuously recorded by the master computer Gaia, and in the event of their demise they can live on in the computer's circuitry—or, should Gaia so decide, in a renewed embodiment. In Frederik Pohl's *The Annals of the*

1. Cited by Carol Zaleski, *The Life of the World to Come: Near-Death Experience and Christian Hope* (New York: Oxford University Press, 1996), p. 10. Apparently the rocket has not yet been launched.

2. Readers interested in cryonics can learn much, much more at: www.syspac.com/~cryoweb/venturism.

3. Varley, *Wizard* (New York: Berkley Publishing, 1980).

Heechee,[4] "machine-stored people" can be successors to, but may also co-exist with, their flesh-and-blood counterparts. The hero, Robinette Broadhead, describes himself as "immortal and yet dead . . . almost omni-scient and nearly omnipotent, and yet no more than a phosphor flicker on a screen."[5] To keep him company, Robinette has "Portable-Essie," a cyber-netic version of his wife; he seldom sees "real Essie," who is still among the living.

Still, it can hardly be denied that these secular versions of immortality have their drawbacks. It may be a source of gratification for the living to think that their ashes will be transformed into sunlight and radiated throughout the universe at the speed of light, but no one can suppose it will mean much to the ashes when the time comes. Aside from the high cost of cryonic suspension,[6] it's obviously a gamble to assume that future science will be able to provide recovery from what has been termed "se-vere, long-term, whole-body frostbite"[7]—as well as from the condition that was the cause of death (or near death) in the first place. Cybernetic immortality not only depends on presently nonexistent technology but faces conceptual difficulties as well. And of course, none of these secular versions of salvation was available to anyone prior to the present genera-tion.

In this final chapter we shall be concerned more particularly with reli-gious hopes concerning "the life of the world to come." These hopes have been a matter of serious belief for large numbers of people over many cen-turies, and they offer the prospect of survival to all human beings rather than to the technologically advantaged few. In order to keep the topic within reasonable bounds, two further limitations must be imposed. First, we shall focus on these hopes as they manifest themselves in the Western theistic religions, primarily Christianity;[8] Eastern views concerning rein-carnation and union with/absorption into the Ultimate Reality will not be explored. Second, we shall not concern ourselves with the *evidence* for

4. Pohl, *The Annals of the Heechee* (New York: Ballantine Books, 1987).

5. Ibid., p. 1.

6. Currently from $28,000 to $150,000, depending on the technology employed.

7. An article in the June 1988 issue of *Cryonics* magazine advocates the view that "cryonics does not actually involve freezing dead people; instead, it redefines death so that cryonic suspension saves terminally ill people, not dead people" (see Kevin Q. Brown, "Death of Death in Cryonics," at http://atropos.c2.net/~kqb/archive/50).

8. The views we will consider are also relevant to Muslims and Jews, but the dis-cussions have actually taken place against a backdrop of Christian belief.

human survival. Instead, our inquiry will be broadly conceptual in nature: we will be discussing the conceivability, coherence, and, to some extent, the plausibility of various views concerning survival, given the well-known facts about the mortality and dissolution of human bodies. These limitations exclude from consideration topics that are deserving of an extensive treatment; even so, we shall have plenty to occupy us. We will proceed by examining the prospects for personal survival after death in the light of several types of views on the mind-body problem.

TRADITIONAL DUALISM AND SURVIVAL

We begin by considering the implications for survival of what may be broadly termed "traditional dualism"—the kinds of dualism that have commonly been assumed in the Western religious traditions. (Platonic-Augustinian dualism, Thomistic dualism, and Cartesian dualism would all qualify here.) One might think that the prospects for survival, given any of these views, should be quite promising; that, after all, is an important part of what enabled them to become "traditional" in the first place. But before we accept a favorable verdict on these dualisms, two types of objections must be confronted: there are theological objections, and also philosophical objections.

The theological objection to dualism is closely related to the ethical objection considered in Chapter 6. Briefly stated, it goes something like this: The biblical writers do not conceive of humans as a compound of two separate substances, but as a single, integrated whole. The terms translated "soul" and "spirit" do not designate discrete, metaphysically separable parts of a person; rather, they are ways of referring to the human personality as a whole. The classical Greek idea of the soul as trapped or "entombed" in the body, and yearning for release into disembodied bliss, is alien to the writers both of the Hebrew Bible and of the New Testament, all of whom regarded the body, and bodily existence, as a good gift from God. The future life to which these writers look forward is an *embodied* life, one that commences at the resurrection and continues on into the age to come. The incorporation of Greek notions of dualism into the Christian tradition was a mistake that seriously compromised the Christian message.

There is an element of truth in this complaint. Unquestionably it is true that the biblical writers view human embodiment as a good thing, even though our bodily nature, like human nature as a whole, is flawed by sin. The ultimate hope is indeed for resurrection rather than for disembodied survival. And the various terms, in Hebrew and Greek, for aspects of the person are fluid and overlapping in their connotations; they cannot without violence be pressed into the mold of a rigid philosophical anthropology.

Nevertheless, these considerations do not justify a blanket rejection of dualism as a model for biblical anthropology. As John W. Cooper has pointed out, the Hebrews clearly had a conception of Sheol as a place where, after death, the person who has died maintains a somewhat shadowy and insubstantial but nevertheless real existence.[9] As Cooper says, "Persons are not merely distinguishable from their earthly bodies, they are separable from them and can continue to exist without them. At death there is a dichotomy of fleshly and personal existence. A person need not be a purely nonbodily substance as in Plato or Descartes for dualism to result. Being an ethereal or quasi-bodily entity will do just as well. The logic is just as inexorable. Dualism is entailed and ontological holism is ruled out."[10] Cooper notes, however, that this type of dualism is compatible with "existential-functional holism" although not with ontological holism. Accordingly, he suggests as a label for the Hebrew view of the person "holistic dualism," a conception that is remarkably similar to Charles Taliaferro's "integrative dualism" as discussed in Chapter 6.

It is, if anything, even more clear that this sort of dualistic conception underlies the anthropology of the New Testament. The general pattern for New Testament eschatology (a pattern already well established in first-century Judaism) involves a three-stage progression: death, followed by a temporary state of disembodied existence, followed by the resurrection and judgment on the last day. Clearly, this involves dualism of a sort, though not necessarily one defined in philosophical categories.[11] It's clear, on the other hand, that the Fathers of the early Christian church did draw upon Greek philosophy in giving an account of the nature of human beings, as well as in delineating the nature of God. But they did not simply baptize the theories of Plato and Plotinus; rather, they "corrected" aspects of the philosophical accounts of persons that came into conflict with

9. John W. Cooper, *Body, Soul, and Life Everlasting: Biblical Anthropology and the Monism-Dualism Debate* (Grand Rapids: Eerdmans, 1989), pp. 77–80.

10. Ibid., p. 77.

11. See ibid., chaps. 4–8.

Christian doctrine.[12] The claim that "traditional dualism" represents a sell-out of biblical conceptions in favor of Greek philosophy really cannot be maintained, though it continues to enjoy an undeserved popularity.

The philosophical objections to traditional dualism amount to a critique of the notion of an immaterial soul. We have no way of recognizing and identifying souls, and therefore no way of counting them. The soul of a person we are talking with could be replaced repeatedly, and we would never notice the difference. The memory-criterion for personal identity, since Locke the one most often invoked, fails because there is no decisive way (unless bodily continuity is assumed) of distinguishing true from false memories. If we conceive of the soul as separated from any body, then, like Wittgenstein's "beetle in a box," it cancels out and drops out of the language-game; we have no significant way of talking about it.

By this time these objections have a faintly archaic flavor about them. They have been discussed at length in the literature and, I believe, have been sufficiently answered.[13] An important point is made by Stephen T. Davis when he complains that, in John Perry's discussion, Perry "seems illicitly to jump back and forth between talk about criteria of personal identity and talk about evidence for personal identity."[14] He goes on to suggest that, although the soul cannot play the role of *evidence* for personal identity (due to the difficulties noted concerning the identification of souls), it may nevertheless provide the *criterion* for such identity.

I think this is correct, but I would put the point in a slightly different way. What needs to be seen is that, in the tradition deriving from Wittgenstein, the word "criterion" is often used in a way that systematically obscures the distinction between metaphysics and epistemology—which of course is hardly surprising, given the antimetaphysical animus of Wittgenstein's thought. Wittgenstein generally abandoned the pursuit of truth-conditions in favor of assertibility-conditions. The result of this is that sentences we have no occasion to *assert* are seen as devoid of *meaning*—and, since assertibility in many cases depends on the existence of

12. This is made clear by Stephen T. Davis in "The Resurrection of the Dead," in Stephen T. Davis, ed., *Death and Afterlife* (New York: St. Martin's Press, 1989), pp. 119–48.

13. See Richard Purtill, "The Intelligibility of Disembodied Survival," *Christian Scholar's Review* 5, no. 1 (1975); Paul Helm, "A Theory of Disembodied Survival and Re-embodied Existence," *Religious Studies* 14, no. 1 (1978); and Davis, "Resurrection of the Dead."

14. Davis, "Resurrection of the Dead," p. 137.

some kind of evidence, meaningfulness and the (at least potential) availability of evidence are closely tied together. There is in all this an implicit verificationism, as well as a rejection of any sort of metaphysical realism: what we can't have evidence for, we can't meaningfully talk about, and the possibility of its existence can't be so much as considered.

From the metaphysical realist perspective that underlies the present discussion, things look very different. An important implication of metaphysical realism is that there may very well exist in reality things of which we humans cannot so much as conceive. There may also exist things we can conceive but for which we lack evidence; our minds—and our language—are simply not the measure of the cosmos. Given this perspective, the metaphysical question about *what it means* for a person to be identical over time is sharply distinct from the epistemological question of *how we can know* this about a particular person.

What, in general, is the form of a metaphysical question about the identity of a substance over time? For such a question to arise, we must have a situation in which, at an earlier time t_1, a certain substance exists, and at a later time t_2 something exists which is somehow derivative from, or a continuation of, or a successor to, the original substance. The question, then, is whether or not changes that have occurred between t_1 and t_2 are sufficient to have ended the existence of the substance that existed at t_1, and brought something else into existence. (Note that what is brought into existence need not itself be a substance; it may, for instance, be a collection of scattered parts.) The criteria for answering this question are essentially implicit in our ordinary concepts of various kinds of objects. These criteria are not necessarily logically complete; they will take account of the sorts of questions that arise with some frequency in practice, but may fail to yield an answer when confronted with highly unusual situations such as those featured in philosophical thought-experiments. And of course, there will often be a measure of vagueness in the criteria. ("Can we really say that this structure, now rebuilt, is the *same building* that existed before the fire, given that only part of one wall was left standing?") One final point to be noted, is that the question is asked about the two stages, *assuming complete knowledge of everything relevant that has transpired in the meantime*. The question is (for example) whether *this building*, with its actual history of destruction by fire, rebuilding, remodeling, and so on, is identical with the building that existed on this site many years ago. The question is metaphysical, not epistemological: if part of the history is not known, the metaphysical question can only be answered hypothetically, *on the assumption of* a certain answer to the remaining evidential questions.

Now, what happens when we apply all this to the question of the identity of (for example) Cartesian souls over time? The extremely interesting result is that there is no question to be answered! The question of identity over time arises only when changes have taken place that might reasonably be thought to have ended the existence of the original substance. But none of the things that can happen to Cartesian souls are such as to call their existence as substances into question, so there is no such question to be answered. Let me repeat: *There is no metaphysical question about the identity of Cartesian souls over time.* This does not mean, as we saw in Chapter 6, that the Cartesian conception of the soul is beyond criticism. But *this particular* criticism simply doesn't get off the ground.

There remains, of course, the epistemological question about how we can identify souls—count them, distinguish them from one another, and so on. But once we have accepted epistemological fallibilism and given up our preoccupation with Cartesian demons, reasonable answers are not too difficult to find. In ordinary circumstances, we count souls by counting living human bodies. Special circumstances may arise—permanent coma, split brains, multiple personalities, and the like—which may cause us to question the "one body, one soul" formula.[15] With regard to disembodied souls, most of us have little or no occasion to identify them at all.[16] But it isn't difficult to think of possibilities. Stephen T. Davis suggests the following: "If there are, say, 100 disembodied souls all wondering whether everyone in fact is who he or she claims to be, it would be irrational to deny that their memories are genuine if they all fit together, confirm each other, and form a coherent picture. Doubt would still be conceivable, but not rational."[17] Friends of Cartesian demons may still not be satisfied— but then, who has ever succeeded in satisfying them about anything?

Let me make this point in another way. Each of the following scenarios seems to be logically possible: (1) The soul of the woman I am talking with has been replaced repeatedly during our conversation, but with no behavioral change that would indicate this to me. (2) The woman has no

15. For a discussion of such cases, see Kathleen V. Wilkes, *Real People: Personal Identity without Thought Experiments* (Oxford: Clarendon Press, 1988).

16. Religious persons do sometimes believe that God is speaking to them or otherwise manifesting his presence to them. Epistemological criteria for identifying the divine presence fall outside the scope of the present work.

17. Davis, "Resurrection of the Dead," p. 139.

subjective consciousness at all; she is really a zombie, though one whose body continues to respond as though she were fully conscious. (3) My conversation partner is in her totality an illusion of mine, as are all the other persons in the world, the entire physical world itself, and my own body; all this is a massive hallucination inflicted on me either by a Cartesian demon or by a mad scientist who has my brain in his vat. (4) I myself have existed for only about five minutes; my memories of the beginning of this very conversation, as well as all the memories of my past life, are false memories implanted in me by the demon/scientist who made me. As I've said, each of these scenarios is logically possible. But with respect to the last three of the possibilities mentioned, most philosophers have learned to take them with equanimity, not considering that they give us reason to give up any beliefs (with the possible exception of beliefs about absolute epistemological certainty) we might otherwise have. Why, then, is the first of the four possibilities still regarded as a source of serious philosophical perplexity?

MATERIALISM AND SURVIVAL: RE-CREATION

Just as it is natural to associate dualism with a belief in survival after death, it is natural to associate materialism with a denial of survival. And in general the association holds good, but not invariably. There exists a group of thinkers who may be termed "Christian materialists"—persons who consider that human beings are entirely composed of ordinary material stuff,[18] yet who embrace the Christian hope for "the resurrection of the dead and the life everlasting." Most of these thinkers affirm resurrection in the sense of *re-creation*: at some time subsequent to their death, persons are *re-created* by God and resume their conscious existence. Christian materialists generally accept, either wholly or in part, the antidualist arguments discussed in the previous section, and they specifically reject the existence of a soul as a "connecting link" between pre-death and post-resurrection existence. A classic statement of the view comes from John Hick, who acknowledges that "a major problem confronting any such doctrine is that of providing criteria of personal identity to link the

18. These thinkers may not (and some clearly do not) endorse the strong versions of materialism discussed in Chapters 2 and 3, involving the causal closure of the physical realm.

earthly life and the resurrection life."[19] As a contribution to this task, Hick offers a series of scenarios:

> Suppose, first, that someone—John Smith—living in the United States were suddenly and inexplicably to disappear before the eyes of his friends, and that at the same moment an exact replica of him were inexplicably to appear in India. The person who appears in India is exactly similar in both physical and mental characteristics to the person who disappeared in America. There is continuity of memory, complete similarity of bodily features including fingerprints, hair and eye coloration, and stomach contents, and also of beliefs, habits, emotions, and mental dispositions. Furthermore, the "John Smith" replica thinks of himself as being the John Smith who disappeared in the United States. After all possible tests have been made and have proved positive, the factors leading his friends to accept "John Smith" as John Smith would surely prevail and would cause them to overlook even his mysterious transference from one continent to another, rather than treat "John Smith," with all of John Smith's memories and other characteristics, as someone other than John Smith.
>
> Suppose, second, that our John Smith, instead of inexplicably disappearing, dies, but that at the moment of his death a "John Smith" replica, again complete with memories and all other characteristics, appears in India. Even with the corpse on our hands, we would, I think, still have to accept this "John Smith" as the John Smith who had died. We would just have to say that he had been miraculously re-created in another place.[20]

This is followed by a third scenario in which the John Smith replica appears, not in India, but "in a different world altogether, a resurrection world inhabited only by resurrected persons."[21]

What shall we say to these stories? For many of us, the first scenario will meet with little resistance: a generation of *Star Trek* viewers will readily grant the conceivability of the instantaneous (or near-instantaneous) transport of persons to a distant location.[22] The heavy work has to be done

19. John Hick, *Philosophy of Religion*, 3d ed. (Englewood Cliffs, N.J.: Prentice-Hall, 1983), p. 125.

20. Ibid., pp. 125–26.

21. Ibid., p. 126.

22. Actually there is a legitimate question whether the methods described in *Star Trek* would result in the transportation of a person, as opposed to the destruction of a

by Hick's second version, in which the corpse is lying before us and yet we are asked to accept the replica in India as the very person who died in America. And even here the case might seem plausible, *if* we were permitted to assume that, upon John's demise, his soul departed from his fleshly remains and now animates the re-created body. But of course we are *not* permitted to assume this; rather, Hick tells us that death is "sheer unqualified extinction—passing out from the lighted circle of life into 'death's dateless night.' "[23] And now John Smith's situation begins to be serious. If John, during his life, is a wholly physical being, then he is, quite simply, a living human body. But that body is still with us, lying dead on the floor—so whatever exactly that thing in India may be, it can't be John Smith; it has to be an impostor of some kind. What the re-creationist must answer is that this body was John Smith *while it was alive*, but now it is John no longer. When John's body perished, John himself passed out of existence—except that, instead of remaining nonexistent, he was re-created by God in a new body.

Opponents of re-creation do not deny that it is within God's power to do what Hick has described. What they say, rather, is that the "replica" described by Hick would be a *mere* replica; it is not conceptually coherent to assert that the replica is the same person who previously lived and died. In support of this, two main lines of argument are given. First, it is claimed that the kind of connection that obtains between the replica and John Smith is not of the right sort to preserve personal identity between them; rather, the scenario amounts to the complete and final destruction of John Smith, and the creation of a replica or reproduction to take his place. Second, it is pointed out that, if it is in God's power to create a single replica, it is equally within his power to create multiple replicas. Clearly several different replicas could not each be identical with John Smith—and, since none has a better claim to that status than any other, the only conclusion to be drawn is that none of the replicas is identical with John. We will need to explore each of these lines of argument in some detail.

Before proceeding, however, some questions need to be clarified about the precise nature of the re-creationist hypothesis. First of all, there is a tension, already noted by Hick,[24] between the apparent requirements of the theory and the nature of resurrection hope. In order to make Hick's

person at one location and the creation of a replica somewhere else. But Trekkies are strongly conditioned to accept the former interpretation.

23. Hick, *Philosophy of Religion*, p. 125.
24. Ibid., p. 126.

scenarios as plausible as possible, the re-created persons are made out to be extremely similar—perhaps exactly similar, apart from the fact of death—to the bodies that have died. But the biblical picture of the resurrection—for example, in I Corinthians 15—is of bodies that have been radically transformed and are unimaginably superior to those we now inhabit. It seems the re-creationist must respond in one of two ways. It may be asserted, in spite of what has just been said, that it is possible for the resurrection body to be sufficiently similar to the body that has died to be the same individual, and yet sufficiently different to satisfy the resurrection hope. Or on the other hand, it may be held that the re-created bodies are initially very similar to the bodies that died, but that, once re-creation has occurred, a process of transformation is begun that preserves personal identity yet results in a glorified resurrection body. For present purposes I will assume that one or the other of these solutions is workable.

Two other questions are also implied in Hick's scenarios. First, there is the *timing* of the resurrection: does it occur immediately after death, as in Hick's stories, or as an event at the end of history, as is suggested by classical Christian eschatology? There is also the question of the *matter* of which the resurrected body is composed: is it the *same matter* as that which composed the body that perished, or can it be new matter entirely, as is implied in Hick's narratives? Interestingly, there is a kind of trade-off between these considerations. An immediate re-creation may seem more likely to be identity-preserving than one that occurs after a long delay, but this requires giving up identity of matter, since it's obvious that the stuff of dead bodies is *not* immediately removed in order to be recycled in a resurrection body. If on the other hand we want to hold out for identity of matter, we will have to allow for a temporal gap—possibly a very long gap—of nonexistence between death and resurrection.

I believe the re-creationist should *not* hold that the matter of the re-created body needs to be the same as that of the body that perished. There is, after all, the very real possibility that some or all of the matter of someone's body might be destroyed (perhaps annihilated in a nuclear reaction) or might become part of someone else's body.[25] Peter van Inwagen puts the point nicely:

25. As a modest addition to the lore on this topic, consider the notion, familiar in science fiction, of a Generation Ship. This is a spacecraft which, in order to cover the vast distances between stars, sets out with the intention of continuing its voyage over hundreds of years, with multiple generations of human beings being born, living, and

If, in order to raise a man on the Day of Judgment, God had to collect the "building blocks"—atoms, neutrons, or what have you—of which that man had once been composed, then a wicked man could hope to escape God's wrath by seeing to it that all his "building blocks" were destroyed. But according to Christian theology, such a hope is senseless. . . . [I]f that theory were true, a wicked man who had read his Aquinas might hope to escape punishment in the age to come by becoming a lifelong cannibal. But again, the possibility of such a hope cannot be admitted by any Christian.[26]

The classical theologians, to be sure, were well aware of this kind of problem, and sought to dispose of it by appealing to divine omnipotence.[27] But even omnipotence cannot incorporate into a resurrection body matter that no longer exists, or fashion two or more simultaneously existing resurrection bodies out of the very same matter. So in order for this requirement to work out, God is going to have to be active in preventing some entirely feasible scenarios from ever taking place.[28] And at this point I suspect that whatever plausibility the view may once have seemed to possess is lost. Apparently the advocates of re-creationism accept this; at least, I know of none who clearly stipulates that the identity-of-matter condition must be met.

---

dying during the journey. Obviously stringent recycling procedures must be followed, so as to avoid exhausting the resources needed to sustain life. It might well then happen that all of the available carbon had, during the course of the voyage, been part of the bodies of dozens or even hundreds of different individuals. Perhaps the best God could do, under these circumstances, would be to resurrect some *very small* persons!

26. Peter van Inwagen, "The Possibility of Resurrection," *International Journal for the Philosophy of Religion* 9 (1978): 114–21; reprinted, with an Author's Note added in 1990, in Paul Edwards, ed., *Immortality* (New York: Macmillan, 1992), pp. 242–46; reference is to p. 245 in the Edwards volume.

27. Peter Geach becomes somewhat belligerent in discussing this point: "The traditional faith . . . is not going to be shaken by inquiries about bodies burned to ashes or eaten by beasts; those who might well suffer just such death in martyrdom were those who were most confident of a glorious reward in the resurrection" (Peter Geach, *God and the Soul* [New York: Schocken Books, 1969], p. 29).

28. Stephen T. Davis suggests, "perhaps omnipotence must . . . guarantee that no essential part of one person's earthly body is ever a constituent part, or an essential constituent part, of someone else's body" ("Resurrection of the Dead," p. 133). Presumably, then, God must ensure that no Generation Ship ever sets out into deep space!

With regard to the timing of the resurrection, it is not clear that one option is philosophically preferable to another. Immediate resurrection may *seem* more likely to conserve personal identity, but it is not at all clear that this is really so. Absent cogent arguments for one side or the other, I propose to take the view that this is a matter that can be settled either way according to considerations of theology and general coherence.

It is somewhat ironic that the most incisive critic of re-creationism is himself a Christian materialist who affirms belief in the resurrection: I am referring to Peter van Inwagen. I am not sure that any of his arguments are entirely without precedent elsewhere, but he has stated them in a vivid and energetic manner that has proved to be convincing to many, including some of his fellow Christian materialists. Van Inwagen's discussion will provide a suitable focus for our consideration of the objections to re-creationism. His most distinctive point is made by means of a story:

> Suppose a certain monastery claims to have in its possession a manuscript written in St. Augustine's own hand. And suppose the monks of this monastery further claim that this manuscript was burned by Arians in the year 457. It would immediately occur to me to ask how *this* manuscript, the one I can touch, could be the very manuscript that was burned in 457. Suppose their answer to this question is that God miraculously recreated Augustine's manuscript in 458. I should respond to this answer as follows: the deed it describes seems quite impossible, even as an accomplishment of omnipotence. God certainly might have created a perfect duplicate of the original manuscript, but it would not be *that* one; its earliest moment of existence would have been after Augustine's death; it would never have known the impress of his hand; it would not have been a part of the furniture of the world when he was alive; and so on.
>
> Now suppose our monks were to reply by simply asserting that the manuscript now in their possession *did* know the impress of Augustine's hand; that it *was* a part of the furniture of the world when the Saint was alive; that when God recreated or restored it, He (as an indispensable component of accomplishing this task) saw to it that the object He produced had all these properties.
>
> I confess I should not know what to make of this. I should have to tell the monks that I did not see how what they believed could *possibly* be true.[29]

29. Van Inwagen, "Possibility of Resurrection," pp. 242–43.

Van Inwagen's little story resonates with several possible objections to re-creationism. It might seem to reflect Locke's dictum that "one thing cannot have two beginnings of existence."[30] A re-creationist who desired to assist the monks might reply that the re-creation of the manuscript by God was *not* the manuscript's "beginning of existence" but merely the resumption of its existence *as a manuscript* after a period either of temporary nonexistence or of existence as scattered parts. The narrative undoubtedly carries overtones of the idea that "essence entails origin";[31] whatever may be the case for other kinds of objects, isn't it clear that the essence *of a manuscript* entails its having originated at the hand of the author? As van Inwagen says, "The manuscript God creates in the story is not the manuscript that was destroyed, since the various atoms that compose the tracings of ink on its surface occupy their present positions not as a result of Augustine's activity but of God's."[32] It isn't easy to think of a convincing reply to this, but perhaps the re-creationist could maintain that those atoms *do* owe their present positions to Augustine's activity: God (it may be said) merely *replaced* the parts of the manuscript after they had been disturbed, much as a mason might replace a bit of stonework that had come dislodged from the building of which it was a part. To be sure, the "parts" of the manuscript have been *completely* separated from each other; it's not just a matter of one or two having been displaced. But as R. T. Herbert points out, there are some objects that clearly can continue to exist despite complete disassembly: for example, a stage set disassembled for removal to the next town, or "illegal weapons completely stripped down for secret transport and later put together again."[33] These, however, are objects that are *constructed* with a view to their being disassembled and subsequently reassembled. (Remember the point that the

30. John Locke, *Essay Concerning Human Understanding*, chap. 27; reprinted in John Perry, ed., *Personal Identity* (Berkeley: University of California Press, 1975), p. 33.

31. Van Inwagen writes: "I should also be willing to defend the following thesis: the thing such an action of God's [viz., the re-creation of a person] would produce would not be a member of our species and would not speak a language or have memories of any sort, though, of course, he—or *it*—would *appear* to have these features" ("Possibility of Resurrection," p. 244). These claims of van Inwagen will not be discussed further here.

32. Van Inwagen, "Possibility of Resurrection," p. 243.

33. R. T. Herbert, "One Short Sleep Past?" *International Journal for Philosophy of Religion* 40 (October 1996): 85–99; the quotation is from p. 92. This article is the best defense of re-creationism against objections I have seen.

criteria for identity over time are kind-specific.)[34] Even more importantly, the "disassembly" stops well short of breaking the objects down to their ultimate constituent atoms: "If the stage set were burned to ashes rather than struck, the . . . weapons melted down rather than disassembled, then their dissolution would end their existence."[35]

In his attempt to find an object that can survive even this radical kind of disassembly, Herbert settles at last on a mythical object: the phoenix. But to meet his requirements, the myth must be interpreted in a particular way. We must understand the phoenix, not as a "series of birds . . . with a unique method of producing offspring," but as a *single individual* such that "however thorough the bird's dissolution in flames, its existence is not terminated." We should not think, however, of the bird's existence as continuing in the form of ashes; instead, "let us think that during its existence the phoenix repeatedly rises to life, leaving countless piles of its ashes like so many abandoned nests."[36] (Note the rejection of the identity-of-matter condition!) And since the bird does not continue to exist as ashes, what we must accept is that *the existence of the phoenix contains temporal gaps.* Just as a performance of a play lasts for a couple of hours, *including* the twenty-minute intermission when nothing is happening on the stage, so the existence of the phoenix continues throughout the interval during which, the phoenix having been incinerated and not yet reconstituted, there is nothing concrete in existence that can be identified as the phoenix.

Given all this, it must be acknowledged that Herbert has constructed an excellent parallel to the theory of resurrection as re-creation. The difficulty, of course, is that all the perplexing features of re-creationism are carried over intact into the phoenix example, with the result that those of us

34. Here I will venture a comment on the Ship of Theseus. I believe the reason why this example is genuinely puzzling is that our ordinary ideas about ships do not contemplate the possibility that an entire ship can be broken down to its parts and then reassembled. Once this possibility is pointed out to us, the claim of the reassembled ship to be the "true" Ship of Theseus has undeniable appeal, and competes (though not decisively) with the claim of the ship that preserved continuity throughout the gradual replacement of parts. If Theseus's vessel had been of a different sort—say, an inflatable that was made to break down into pieces for portaging—then I for one would be much more ready to say that the reassembled boat was the "true" Ship of Theseus.

35. Herbert, "One Short Sleep Past?" p. 92.

36. Ibid., p. 93.

who were baffled to begin with are likely to remain baffled still. The reader must decide for herself what to make of all this.

It remains to consider the multiple-replica objection. Van Inwagen phrases the objection like this:

> If God can, a thousand years from now, reassemble the atoms that are going to compose me at the moment of my death—and no doubt He can—, He can also reassemble the atoms that compose me right now. In fact, if there is no overlap between the two, He could do both, and set the two resulting persons side by side. And which would be I? Neither or both, it would seem, and since not both, neither.

> "God wouldn't do that." I dare say he wouldn't. But if He were to reassemble either set of atoms, the resulting man would be who he was, and it is absurd, it is utterly incoherent, to suppose that his identity could depend on what might happen to some atoms other than the atoms that compose him.[37]

(It will be noted that van Inwagen's phrasing of this objection assumes the identity-of-matter requirement. But the objection doesn't depend on that; it's even more obvious that multiple replicas are possible if this requirement doesn't obtain.)

Re-creationists respond to this objection in various ways. Stephen Davis quotes with approval a line from a dialogue by John Perry, in which a character says to his dying friend, "Why can't Sam simply say that if God makes one such creature, she is you, while if he makes more, none of them is you? It's possible that he makes only one. So it's possible that you survive."[38] The reason this can't work is that identity is a relation that, if it holds at all, holds necessarily.[39] As van Inwagen says, it really is

37. Van Inwagen, "Dualism and Materialism: Athens and Jerusalem?" *Faith and Philosophy* 12 (October 1995): 486.

38. From John Perry, *A Dialogue on Personal Identity and Immortality* (Indianapolis: Hackett, 1978); quoted in Davis, "Resurrection of the Dead," p. 139. It will be noted that Davis defends the coherence both of traditional dualism and of re-creationism; I take it he presently accepts some form of dualism.

39. Suppose you are attending a meeting of the American Philosophical Association and someone comes up and asks whether you are Alvin Plantinga. You reply, "I might be, but I'm not." Wouldn't this have to be a joke?

absurd to suppose that a person's identity could depend on what might happen to some atoms other than the atoms that compose him or her.[40] An elegant formal proof of the necessity of identity is given by J. J. MacIntosh:

(1) $(a = b) \rightarrow (\Box (a = a) \leftrightarrow \Box (a = b))$ Leibniz's Law
(2) $a = a$ Reflexivity of Identity
(3) $\Box (a = a)$ 2, and the Rule of Necessitation
     (all theorems are necessarily true)
(4) $(a = b) \rightarrow \Box (a = b)$ 1, 3

Equivalently,

(5) $\Diamond (a \neq b) \rightarrow (a \neq b)$ 4, Contraposition, and the definition of $\Diamond$

This says that if a and b are *possibly* nonidentical, then they *are* nonidentical, which is just what is needed to let van Inwagen's argument go through.[41]

Herbert answers the argument from the necessity of identity as follows:

> Perhaps it is enough of a reply to this argument to point out that its modal mainspring has amusing applications. For example, the woman beside whom I awakened in bed this morning is not my wife, since she is possibly not my wife, since it is possible that during the night my wife split (I mean, underwent "mitosis"), resulting in two equally well qualified contenders for my-wifehood. Fortunately for me she did not undergo this process, for

---

40. Harold Noonan has pointed out that there is a way in which one can hold a "closest continuer" theory of personal identity without denying the necessity of identity. But this alternative has other consequences which are, if anything, even more unpalatable than denying the necessity of identity. For details, see Noonan's *Personal Identity* (London: Routledge, 1989), chap. 7.

41. See J. J. MacIntosh, "Reincarnation and Relativized Identity," *Religious Studies* 25 (1989): 158. Unfortunately, MacIntosh proceeds to interpret possibility, as represented by the '$\Diamond$' in (5), as *epistemic* possibility. This is a blunder: identity is not an epistemic relation, and the possibility (or even the actuality) of an *epistemic* contender for an identity-claim showing up does *not* invalidate the claim. By making this mistake, MacIntosh sets himself up for Herbert's (equally misguided) rejoinder, cited below in the text.

I am in rather late middle age, beyond the conduct of a *menage a trois*.
Even so, it is distressing to learn that this woman is not my wife because
she is not necessarily my wife.[42]

This is indeed amusing, but I am afraid the joke is on Herbert. The problem is that he (like a number of other philosophers) has failed to understand the proper form of a question about personal identity over time. As
we saw in the last section, the question is not whether something conceivably could have happened that would have had the result that the woman
by his side was nonidentical with the woman Herbert married. The question is whether, *consistent with her actual history*, there could be some
other woman with an equally strong claim to be identical with Herbert's
bride. And of course the answer to this is No. She *has not*, as a matter of
fact, undergone mitosis, or annihilation and re-creation; she remains simply herself, and anyone else claiming to be Herbert's bride has to be an
impostor. She is not, to be sure, *necessarily* Herbert's wife; being his wife is
not an essential property of hers. But she is necessarily identical with the
woman who is in fact Herbert's wife, and that ought to be good enough
for all concerned.

The problem for re-creationism, on the other hand, is that, under the
"multiple replica" scenario, there could very well be multiple contenders,
each with an equal claim to be identical with the person who died. In that
case, clearly, none of the replicas would be identical with the dead person.
But since it is possible that there be multiple replicas, it is possible that the
re-created person is nonidentical with the deceased—and, to return to the
point, if they are possibly nonidentical, they *are* nonidentical, and indeed
necessarily nonidentical.

At this point readers are invited to decide for themselves how to assess
these arguments. None of the antimaterialist arguments of the preceding
chapters depend on the points at issue here, so in principle I could accept
either verdict about the coherence of re-creationism. I must confess, however, that I think it will be exceedingly difficult for the re-creationist to
argue either that the monks are right about the Augustine manuscript or
that the re-creation of persons differs from that case in some way that
would justify a contrary verdict. Furthermore, the only way to defeat the

---

42. Herbert, "One Short Sleep Past?" p. 97.

"multiple replica" objection is to deny the necessity of identity—and once that has been done, we are no longer talking about *identity*.

MATERIALISM AND SURVIVAL: BODY-SWITCHING AND BODY-SPLITTING

Van Inwagen's critique of re-creationism is powerful, but van Inwagen himself affirms the possibility of resurrection. He hasn't made things easy for himself, however. He clearly is committed to each of the following propositions:

1. A human being is a concrete individual, not a group of objects or an abstract object such as a computer program.
2. The very same human beings that live and die in our present world will live again in the resurrection.
3. There is no nonmaterial part of a human being that survives bodily death.
4. In order for a person who has undergone death and dissolution to live again, it is not sufficient that all of the atoms of the body be reassembled in their exact relative positions and caused to resume their vital functions; a fortiori, it is not sufficient to produce an exact replica using different atoms.

Given these assumptions, the prospects for survival do not seem bright. As van Inwagen says, "I conclude that my initial judgment is correct and that it is absolutely impossible, even as an accomplishment of God, that a man who has been burned to ashes or been eaten by worms should ever live again. What follows from this about the Christian hope of resurrection? Very little of any interest, I think. All that follows is that if Christianity is true, then what I earlier called 'certain facts about the present age' are *not* facts."[43] And now van Inwagen is ready to make his case for the possibility of resurrection:

43. Van Inwagen, "Possibility of Resurrection," p. 245. One may perhaps be permitted the comment that van Inwagen here shows himself a person whose interest is hard to engage.

It is part of the Christian faith that all men who share in the sin of Adam must die. What does it mean to say that I must die? Just this: that one day I shall be composed entirely of non-living matter; that is, I shall be a corpse. It is not part of the Christian faith that I must at any time be totally annihilated or disintegrate. . . . It is of course true that men apparently cease to exist, those who are cremated, for example. But it contradicts nothing in the creeds to suppose that this is not what really happens, and that God preserves our corpses contrary to all appearance. . . . Perhaps at the moment of each man's death, God removes his corpse and replaces it with a simulacrum which is what is burned or rots. Or perhaps God is not quite so wholesale as this: perhaps He removes for "safekeeping" only the "core person"—the brain and central nervous system—or even some special part of it. These are details.[44]

I think van Inwagen has made a good case for the logical possibility of resurrection. The removal by God of the dead person to some unknown location—whether in our space-time continuum or outside of it—is surely something a theist must affirm to be possible. It is true that the story requires that vital functions cease and are later "jump-started" by divine power, but I think it is at least very plausible to suppose that the person's identity can be maintained throughout such a process. (Certainly we show little inclination to question the identity of persons whose hearts are restarted by artificial means after they are "clinically dead.") There is nothing else in the story that even seems to create conceptual difficulties.

Van Inwagen, however, affirms not only the possibility of resurrection but its actuality.[45] And as an account of the way in which God actually raises the dead, his narrative leaves much to be desired. There is an intriguing similarity between van Inwagen's story and the scenario put forward by cryonics. At death, God plays the part of the practitioners of cryonics, placing the body in suspension to prevent further damage or deterioration. Then at the resurrection, God assumes the role of the future medical rescuers: he reanimates the corpse, heals its fatal injury or ill-

44. Ibid., pp. 245–46.

45. The aim of his 1978 article is merely to demonstrate logical possibility, and in this I think he succeeds. But in 1982 van Inwagen became a Christian (see his autobiographical essay, "Quam Dilecta," in Thomas V. Morris, ed., *God and the Philosophers: The Reconciliation of Faith and Reason* [New York: Oxford, 1994], pp. 31–60), and a believer in an *actual* resurrection.

ness, and puts the revitalized person on the road to a fuller life. (It is left to the reader to decide whether this similarity is a strength or a weakness of van Inwagen's view!) But his view diverges from the cryonics scenario with respect to the massive deception God is practicing if the bodies we inter are not really the bodies of the persons who have died. Dean Zimmerman recounts that he once assisted a friend who was an anatomy student in dissecting a corpse. He observes that

> Opening a human skull and finding a dead brain is sort of like opening the ground and finding a dinosaur skeleton. Of course it is in some sense possible that God takes our brains when we die and replaces them with stuff that looks for all the world like dead brains, just as it is possible that God created the world 6000 years ago and put dinosaur bones in the ground to test our faith in a slavishly literal reading of Genesis. But neither is particularly satisfying as a picture of how God actually does business.[46]

It seems, in fact, that this feature of van Inwagen's story violates stipulations van Inwagen himself has elsewhere accepted as constraints on acceptable solutions for theological problems. In his essay "The Problem of Evil, the Problem of Air, and the Problem of Silence" van Inwagen asserts that an acceptable "defense" against the problem of evil must consist of propositions which, though they perhaps cannot be shown to be true, are nevertheless "true for all anyone knows."[47] I want to say that we know just as well that the object that decays after death is the body that once was alive, as we know that any other physical object enjoys continuous existence. In that same essay, van Inwagen sets out the claim that "being massively irregular is a defect in a world,"[48] a defect a great as the defect of containing suffering on the scale that exists in the actual world. A world in which systematic body-switching occurs would seem to be "massively irregular" if any world deserves that description. And what would our at-

46. Dean A. Zimmerman, "The Compatibility of Materialism and Survival: The 'Falling Elevator' Model," *Faith and Philosophy*, forthcoming, sec. 1.

47. Van Inwagen, "The Problem of Evil, the Problem of Air, and the Problem of Silence," in *God, Knowledge, and Mystery: Essays in Philosophical Theology* (Ithaca: Cornell University Press, 1995), p. 75.

48. Ibid., p. 76. It should perhaps be pointed out that this proposition is itself part of a story "that is true for all I [van Inwagen] know," and so it is not actually asserted as true. Still, it is hardly desirable to have "defenses" for two theological problems that contradict each other!

titude be toward the "remains" of the deceased if we seriously believed van Inwagen's story to be the truth?

It would seem that van Inwagen is not entirely insensible of these difficulties. In the 1990 "Author's Note" appended to his article, he states, "If I were writing a paper on this topic today, I should not make the definite statement 'I think this is the *only* way such a being could accomplish it [viz., resurrection].' I am now inclined to think that there may well be other ways, ways that I am unable even to form an idea of because I lack the conceptual resources to do so."[49] But in a more recent article, he still affirms that "when I die, the power of God will somehow preserve something of my present being, a *gumnos kókkos*,[50] which will continue to exist throughout the interval between my death and my resurrection and will, at the general resurrection, be clothed in a festal garment of new flesh."[51]

More recently, Dean Zimmerman and Kevin Corcoran have attempted to come to van Inwagen's assistance by providing an alternative view that is consistent with his original assumptions.[52] Their developments of this view differ in details; I'll begin by presenting a relatively simple statement of it by Corcoran, and will then add elaborations as they become necessary.

The key idea of the new model can be stated simply: Instead of body-switching (as in van Inwagen's model), we have body-splitting. Corcoran states the matter thus:

49. In Edwards, *Immortality*, p. 246.
50. "Bare kernel"—see I Corinthians 15:37.
51. Van Inwagen, "Dualism and Materialism," p. 486.
52. See Zimmerman, "Compatibility of Materialism and Survival"; and Kevin Corcoran, "Persons and Bodies," *Faith and Philosophy*, forthcoming. Zimmerman's and Corcoran's respective attitudes toward the proposed model differ in an interesting way. Corcoran connects the model with a version of a "constitution" account of human persons, and he puts it forward as a serious candidate for acceptance. Zimmerman, on the other hand, is a dualist who does *not* embrace the model as his own. His objective, however, is to argue that a *materialist* has no good reason not to embrace it, since (he argues) all of the more questionable assumptions of the model are assumptions to which materialists (in particular, van Inwagen) are already committed. In proposing an earlier version of the model, he wrote, "I offer Peter this 'just so story,' to do with as he will, with my compliments. I'm glad I'm a dualist with less need of it" (Comment on van Inwagen's "Dualism and Materialism," delivered at the University of Notre Dame, November 3, 1994).

> Suppose the simples composing my body just before my death are made by God to undergo fission such that the simples composing my body then are causally related to two different, spatially segregated sets of simples. Let us suppose both are configured just as their common spatiotemporal ancestor. Suppose now that milliseconds after the fission one of the two sets of simples ceases to constitute a life and comes instead to compose a corpse, while the other either continues on in heaven or continues on in some intermediate state. It looks to me like the defender of constitution has got all she needs in order to make a case for my continued existence, *post mortem*. For according to this story, the set of simples that at one time composed my constituting body stands in the right sort of causal relation—the Life-preserving causal relation—to the set of simples that either now compose my constituting body in heaven or compose my constituting body in an intermediate state.[53]

Corcoran goes on to say, "While I do not claim that this is the only way of making sense of resurrection on a constitution account of human persons it does seem to be one plausible way of doing so. Moreover, this materialist account of resurrection, unlike van Inwagen's, does not involve God in mass deception. The corpse composed of the set of simples that fails to perpetuate a life is no simulacrum—it really is the stuff that at one time constituted the person in question."[54]

In order for this model to work, we must assume, according to Zimmerman, that God can bring about the fission of an organism by "just before it completely loses its living form, enabling each particle to divide—or at least to be immanent-causally responsible for two resulting particle stages."[55] Zimmerman here mentions two distinct ways in which the organism's fission might take place. It might be, that each simple particle of the original body divides into two different particles, one of which becomes part of the corpse and the other of the resurrection body. But (according to Zimmerman and Corcoran) that is not the only way fission could happen. What is essential, according to them, in order for the resurrection body to be the same as the body that dies, is that there should be appropriate *immanent-causal connections* between the two. Zimmerman gives a rough statement of this requirement when he says, "for an object that persists throughout a given period of time, the way the object is at

53. Corcoran, "Persons and Bodies," sec. 6.
54. Ibid.
55. Zimmerman, "Compatibility of Materialism and Survival," sec. 3.

any moment in that interval must be partially determined by the way it was during the interval leading up to that moment."⁵⁶ Zimmerman is at pains to argue that the requirement "does not rule out the possibility of discontinuous spatiotemporal jumps for objects, or even of 'temporally gappy' objects; it merely describes a condition that applies to periods of time *throughout which* an object exists."⁵⁷

Given all this, we can now consider more closely what occurs at the moment of fission. Zimmerman explains the point as follows:

> If the ultimate simples in my body are the kinds of things that can last through time . . . it will turn out that each simple which God "zaps" with this replicating power in fact does not *itself* divide, but simply remains right here—as a part of my corpse. Each particle *x* is immanent-causally connected to two streams of later particle-stages; one of them—the one in the here and now—includes stages of *x* itself; the other, the one in the here-after, consists of stages of a different particle.⁵⁸

So we have the following situation: the *particles* of the original body become part of the corpse, thus vindicating Corcoran's claim that "the corpse composed of the set of simples that fails to perpetuate a life is no simulacrum—it really is the stuff that at one time constituted the person in question." Zimmerman, however, argues that "the diversity of the particles I'll have after death from the particles in my dying body does not . . . prevent the *bodies* from being the same. All that matters for the continuation of my Life is that the right kind of life-sustaining causal continuity obtain among person-stages." (Note that this sort of "life-sustaining causal continuity" does *not* obtain between the dying body and the corpse, since life in fact is not sustained.)

One point that remains to be considered is the precise moment at which fission occurs. As we've seen, Corcoran proposed that it takes place milliseconds before the person's death. Zimmerman, however, takes a slightly different line, and Corcoran now agrees with this.⁵⁹ According to Zimmerman, fission takes place at the very last instant of my Life—the instant such that at any later instant I would no longer be alive. One ben-

56. Ibid., sec. 3. For a more detailed treatment, see Zimmerman's "Immanent Causation," *Philosophical Perspectives* 11 (1997).

57. Zimmerman, "Compatibility of Materialism and Survival," sec. 3.

58. Ibid.

59. Private communication from Kevin Corcoran.

efit of this move is that, on Zimmerman's proposal, there is no period of time whatsoever (not even a few milliseconds) during which there are two living bodies which might be "equal claimants" to identity with the dying person. And now we are ready for Zimmerman's summary:

> Thus we have a story that includes everything we want: The heap of dead matter I leave behind is made of stuff which really was a part of my body (it is not a simulacrum: God is not a body-snatcher), and the resurrected body is really identical with this present one—it is causally continuous with it in just the way adjoining stages of my present body are causally continuous, except that in this case there is a spatial or spatiotemporal gap which my poor body was given the power to cross by means of God's intervention.[60]

As should already be clear, there is a metaphysical price tag for accepting this proposal. Part of the price involves giving up the assumption, which we normally take for granted, that *continuity of matter* is a requirement for the identity of physical objects over time. In some cases—in particular, living organisms—we allow for the gradual replacement of some or all of the matter of which an object is composed, but we do *not* normally allow for all-at-once replacement, such as is said to take place in the case of the resurrection body. Readers will have to decide for themselves whether it is acceptable to give up the continuity-of-matter requirement in favor of the requirement of immanent-causal connections between body-stages.

Another part of the price to be paid for the Zimmerman-Corcoran model is that it requires us to accept a "closest continuer theory" for the survival of persons. They recognize that van Inwagen would reject this; he has said, after all, that "it is absurd, it is utterly incoherent, to suppose that [a man's] identity could depend on what might happen to some atoms other than the atoms that compose him." Zimmerman and Corcoran, however, think they have an argument that can force him to accept it. The argument goes like this: Suppose it is possible for a human being to be divided into two parts, each of which, given appropriate medical support, is able to survive and function as a person.[61] Let's call the original person Mark, and the two halves into which he is divided $Mark_a$ and $Mark_b$. If both $Mark_a$ and

60. Zimmerman, "Compatibility of Materialism and Survival," sec. 3.

61. Zimmerman's version of the argument, which is by far the more fully developed, is conducted in terms of a mythical creature, van Inwagen's "Neocerberus." But for our purposes it will be simpler to stick with human beings. Note, furthermore, that it is left open *how much* of the original body would have to be "split" in order for each part to continue to exist as a person.

Mark$_b$ survive, Mark's life is ended, though we should hardly want to say that Mark has *died*. Perhaps what we should say is, that Mark's life *as a single individual* has ended, but that his life continues in that of Mark's "successors," Mark$_a$ and Mark$_b$. (Note that since we are assuming materialism, we cannot say that Mark is whichever of Mark$_a$ and Mark$_b$ has Mark's *soul*.)

Suppose, on the other hand, that instead of Mark$_b$ being surgically separated from Mark$_a$, the cells that would constitute Mark$_b$ are *destroyed*—by laser surgery, for example, or by an accident. If what is left is still able to live on and to function as a person, the natural thing to say—the thing all of us will want to say—is that Mark has *survived* the loss of half his tissue. Yet there is *no difference* between the cells that, in this scenario, constitute Mark and those which in the previous scenario constituted Mark$_a$, Mark in that case having failed to survive. Zimmerman concludes:

> I am convinced that *any* materialism concerning human beings that eschews temporal parts can be driven in similar fashion toward a closest continuer account of human persistence conditions. Such materialists cannot avoid saying that, if there are two simultaneously existing and equally good candidates for being involved in the same Life as some earlier person, then the original person ceases to exist; the Life ends, and two new Lives begin. But if one of the two candidates had been completely absent (destroyed at the point of fission instead of being preserved alive), then the original Life would have continued and the original person would have persisted through the loss of half her brain.[62]

What resources does van Inwagen have that might enable him to resist this argument? I propose, first of all, that we accept as data Zimmerman's conclusions about the results of the two thought-experiments. That is to say: If both Mark$_a$ and Mark$_b$ survive, Mark's life is ended, whereas if the cells that would constitute Mark$_b$ are destroyed, Mark's life can continue. Is there any way van Inwagen can account for this, without accepting a closest continuer theory?

Suppose we adopt the view of identity over time sketched in the previous sections of this chapter and apply it to our thought-experiments.[63]

62. Zimmerman, "Compatibility of Materialism and Survival," sec. 2.
63. It should be noted, however, that the line proposed here is not the line van Inwagen actually takes in dealing with such cases. (See van Inwagen's *Material Beings* [Ithaca: Cornell University Press, 1990], pp. 202–12.) Against van Inwagen's position, Zimmerman's argument appears to be successful.

The question then becomes: given the *actual history* of the surviving individual, is it *possible* that there should be an "equal claimant" to identity with Mark, the person who lived previously? In the first thought-experiment, the answer is clearly Yes—in fact, there is an *actual* equal claimant, namely Mark$_b$. So Mark has not survived.

But what about the second version? Isn't it possible, in that case, that Mark$_b$ should have survived as well as Mark$_a$, so that we would be forced to say that here, also, Mark has failed to survive? But we've already agreed that in this case Mark *does* survive, so it looks as though our theory is incompatible with our data.

To see why this is a mistake, recall that the question to be asked is whether it is *consistent with the actual history* of the surviving individual that there should be another "equal claimant" to identity with the person in the past. And in our second thought-experiment, the answer to this question is No. For in that experiment, the destruction of the cells that would constitute Mark$_b$ is *an event in Mark's own life*. Mark's history includes the fact that approximately one-half of his cells were destroyed by an operation or an accident. And given that these cells were in fact destroyed, there is no way they could constitute a competitor for Mark's identity. So our theory is after all consistent with the data we have agreed to accept.

We now see that van Inwagen has a choice of two theories by which to account for the data—the closest continuer theory and the alternative developed here. Which should he choose? The answer is clear: van Inwagen ought to prefer, and we all ought to prefer, the latter theory, the one that affirms that the persistence of a person is negated by the mere *possibility* of an equal competitor. And the reason this theory is preferable is also clear: it enables us to retain classical identity as a relation that holds for persons over time. An "identity relation" that is merely contingent is not *identity*,[64] and to accept a closest continuer theory for the persistence of persons is in effect to admit that no person is identical with a person that existed at an earlier period of her own life. And this is a price none of us should be willing to pay.[65]

---

64. Once again, let me acknowledge that Harold Noonan has shown that there can be a version of the closest continuer theory that avoids making identity a contingent relation. The price of this, however, is accepting other assumptions that are at least equally problematic. For details, see the reference given in note 40.

65. Apparently Zimmerman would agree with this, when speaking in his own,

But if this conclusion holds, the consequences for the Zimmerman-Corcoran proposal are grim. Let's term the "fission products" in their story $Mark_\alpha$ (the resurrected person) and $Mark_\beta$ (the corpse), so as to distinguish them from the surgically separated $Mark_a$ and $Mark_b$. On the present view concerning the persistence of persons through time, it is clear that $Mark_\alpha$ is *not* Mark, and Mark does not survive the fission. One might be tempted to think otherwise. After all, the final result in the fission case parallels that in our second thought-experiment, in which Mark is admitted to survive. In one case the tissues that would constitute $Mark_b$ are destroyed and their elements dispersed; in the other this is what happens (eventually) to the tissues constituting $Mark_\beta$. Quite so—but it makes a difference *how* those tissues were destroyed. In the second thought-experiment, the destruction of those tissues was *part of Mark's life-story*, but this is not so in the fission case. In that case, the tissues are destroyed only *after* fission has occurred—and necessarily so, since prior to the fission there is only *one* set of tissues that constitute Mark. So it is possible for there to be, after the fission, an *actual equal claimant* to identity with Mark, for it is possible, consistent with everything that occurs in Mark's life-story, that the tissues that constitute $Mark_\beta$ might continue in a living state, instead of perishing as in fact they do. So it is entirely possible that this equal claimant—namely, $Mark_\beta$—should have survived along with $Mark_\alpha$. (I believe, in fact, that we should have to say, in that case, that $Mark_\beta$ *rather than* $Mark_\alpha$ is Mark—for $Mark_\beta$ satisfies the continuity-of-matter requirement, and $Mark_\alpha$ does not.) The conclusion is inexorable: Since it is *possible* that $Mark_\alpha$ is *not* Mark, it's also true that *in fact* $Mark_\alpha$ is not Mark; indeed, he *necessarily* fails to be Mark.

I conclude, then, that Zimmerman and Corcoran have failed in their attempt to provide us with an account, alternative to van Inwagen's, of the survival of persons that are purely material. Combining this with the earlier argument against re-creationism, we arrive at the following result: *If you are a materialist who wants to affirm survival, van Inwagen's body-switching scenario is still the only game in town.*

---

nonmaterialist voice. He writes, "Some will insist that adopting a closest continuer theory of personal identity is just as wildly implausible as supposing that God is a body-snatcher—and, for the record, I am inclined to agree" ("Compatibility of Materialism and Survival," sec. 1).

EMERGENT DUALISM AND SURVIVAL

After threading our way through the maze of materialistic versions of survival, it is something of a relief to turn to the theory that represents this book's major proposal, namely emergent dualism. Many of the issues that could arise concerning emergent dualism and survival have already been anticipated in the section on traditional dualism, but something more needs to be said. The three main questions are: Can the emergent self survive? Can it remain identical over time? And can it be reembodied in the resurrection?

In principle, emergent dualism leaves open the question of life after death for human beings. Certainly the theory provides no metaphysical guarantee of survival. If anything, the "magnetic field" analogy cuts the other way: stop the generator, destroy the magnet, and the field disappears. It seems clear, however, that there is at least the logical possibility for the field to continue without its supporting magnet; it is, after all, a distinct individual. An intriguing parallel comes from the physical theory of "black holes." According to Roger Penrose, "After the body has collapsed in, it is better to think of the black hole as a self-sustaining gravitational field in its own right. It has no further use for the body which originally built it!"[66] This quotation is remarkable in showing the possibility, even within physics, that a field generated by a material object could persist in that object's absence.[67] But as an analogy for life after death, it is less than appealing: it suggests the notion of an "elite corps" of high-powered souls that possess the internal energy to become independent of their bodies, while others lack the resources to do this and either perish imme-

66. Roger Penrose, "Black Holes," in *Cosmology Now* (New York: Taplinger, 1976), p. 124. For a more recent statement, consider the following from Kip Thorne: "The singularity and the stellar matter locked up in it are hidden by the hole's horizon. However long you may wait, the locked-up matter can never reemerge. The hole's gravity prevents it. . . . For all practical purposes, it is completely gone from our Universe. The only thing left behind is its intense gravitational pull" (*Black Holes and Time Warps: Einstein's Outrageous Legacy* [New York: Norton, 1994], p. 30). But Thorne also says, "We do not yet know for certain what inhabits a hole's core" (p. 450).

67. Another example of this possibility is found in the discovery, by the theoretical physicist Mael Melvin, that a sufficiently intense magnetic field can hold itself together by gravity even if its generating magnet has been removed. See Thorne, *Black Holes and Time Warps*, p. 263.

diately at the time of death or slowly dissipate thereafter. Much more attractive is the suggestion made by the neuroscientist Wilder Penfield, who hypothesized that throughout life the mind is supplied with energy by the brain, but says, "Whether there is such a thing as communication between man and God and whether energy can come to the mind of man from an outside source after his death is for each individual to decide for himself. Science has no such answers."[68] Indeed it does not. But there is little doubt that an omnipotent God could, for example, annihilate all of the electromagnets in a particle accelerator, and instantaneously replace them with others, while causing the identical field to persist in being. Alternatively, he could directly sustain the field by his own power, without the need for a material "generator" of any kind.[69] Both these scenarios model possible ways in which God could sustain the lives of human persons after their biological death. It seems clear, then, that emergent dualism is far better placed than any kind of materialism in accounting for the survival of persons.

But is the emergent soul-field, as described in the previous chapter, the sort of thing that can possess self-identity over time? Clearly, the answer assumed here is Yes; are there any serious reasons for thinking otherwise? One might suppose that, were God to choose to maintain an emergent self in existence, he would be sufficiently astute to make sure that it was *the same* self he was maintaining, rather than a series of distinct individuals! One question that is bound to arise is that of memories in a disembodied state. It's clear that, in our embodied lives, memories are somehow stored in the brain, but a disembodied self lacking memories of its previous life would be crippled at best. Emergent dualism, then, must postulate either that the disembodied self somehow carries its memories with it apart from their material embodiment in the brain, or that God, as he sustains the existence of the self, also provides it with the relevant memo-

68. Wilder Penfield, *The Mystery of the Mind* (Princeton: Princeton University Press, 1975), p. 215.

69. Kevin Corcoran has suggested to me that perhaps God *could not* do either of these things, on the ground that identity of the generating body is essential for the identity of the field. I don't think there is a compelling case for this, though there is some intuitive pull in this direction. Clearly, I am committed to denying that the continuing identity of the self depends on that of its body. If we were to conclude that the identity of the magnetic field does depend on that of its generating body, that analogy would lose a good deal of its attractiveness.

ries.[70] Either alternative seems possible; there is no need at this point to decide between them.

Other questions arise concerning the continuity and identity of the self in its ordinary, embodied state. In recognizing the fundamental dependence of the self on its brain and body, emergent dualism opens itself to questions about personal identity to which Cartesian dualism is immune. A full discussion of these questions lies outside the present project, but a few things can be said. The division of consciousness that occurs in commissurotomy seems to be partial and temporary: only some aspects of personal functioning are affected, and over time the brain seems to develop alternative channels of information exchange by which to reintegrate the cerebral hemispheres.[71] Cases of multiple personality may be more intractable, though here also reintegration is often possible. Both these types of cases stop somewhere short of complete fission. But it is conceivable—whether or not it is actually possible—that true fission might occur, for instance as the result of surgery in which each of the two cerebral hemispheres survives in a different body. If that were to happen, we would have two persons who were "successors" to the original, undivided personal life; each would have veridical memories of what had been done and suffered in that life. In that case, presumably we should have to make some arrangement whereby the assets and liabilities of the previous life were divided equitably between the successors. One may perhaps be permitted the hope that such contingencies remain hypothetical!

Suppose, finally, that the brain and nervous system of a living body were to enter a state of suspension in which the generation of the conscious field stops altogether. This might be the result of a profound coma or the cryonic suspension of a still-living body. If the cessation is irreversible, it may be plausible to identify this as the moment of death. But if reversal is possible, we should want to say that, during the cessation, the field has a "virtual existence" in the physical system which has supported it in the past and may do so again.[72] Were reversal actually to occur, there

70. Some such assistance would be needed in any case, if the self is to remember the many things in its past for which the memory traces have been extirpated from the brain.

71. Kathleen V. Wilkes contends that the commissurotomy results create less difficulty for our conception of a unified self than do such familiar phenomena as weakness of will and self-deception. See her *Real People*, chap. 5.

72. It's also conceivable that during the interval the field of consciousness might still exist, apart from its material generator. But speculating about this would take us too far afield.

is little doubt we would in practice acknowledge the same person to be in existence after the hiatus.

If survival is possible, then so, it would seem, is resurrection. If God can sustain the field of consciousness absent any material "base" whatsoever, then God can also provide a new base in the form of a resurrection body. This body, to be sure, would have to be crafted specifically to support and energize the particular conscious field in question; in view of the intimate dependence of the mind on the brain, not just any body and brain would do. And this means that the prince-and-cobbler type of body-switching, familiar to us from dualist thought-experiments, really is not possible on this view. One question that has been raised is this: Wouldn't the newly formed resurrection body generate its *own* field of consciousness, and thus be unavailable to the self in need of reembodiment?[73] It seems that this would indeed be the case, if we suppose that the body is first created, with its vital functions energized, *before* the "infusion" of the disembodied self. Rather, we must imagine the new body created from the very beginning *as the body of this very soul*; the renewed self must be "in charge" of the resurrection body right from the start.

No doubt all of this is at best a "likely tale." We need not suppose we have plumbed the secrets of the hereafter, but we may hope to have shown that no insuperable difficulties, either empirical or conceptual, stand in the way of the general account of things that has been offered here. If even that much has been accomplished, there is cause to rejoice. We humans do, after all, have a deep need somehow both to conceive and to imagine for ourselves that which, in its reality, must lie beyond all our concepts and imaginings. And surely it is right that we should do this. In the words of Carol Zaleski, "We have every right to fit ourselves in advance for a pair of wings as long as we recognize that such expectations are no proper measure of the surprise that awaits us."[74] Long before, the Apostle Paul wrote, "When the perishable has been clothed with the imperishable, and the mortal with immortality, then the saying that is written will come true: 'Death has been swallowed up in victory.' "[75]

73. I owe this question to Edwin Hui.
74. Zaleski, *Life of the World to Come*, p. 3.
75. I Corinthians 15:54 (NIV).

Index

237